T0244575

SHIP OF LOST SOULS

The Tragic Wreck of the Steamship Valencia

ROD SCHER

FOREWORD BY CDR. JOHN E. HARRINGTON, USCG (RET.)

ILLUSTRATIONS BY MOLLY DUMAS

ESSEX, CONNECTICUT

An imprint of The Globe Pequot
Publishing Group, Inc.
64 South Main Street
Essex, CT 06426
www.globepequot.com

Distributed by NATIONAL BOOK NETWORK

British Library Cataloguing in Publication Information available
Library of Congress Cataloging-in-Publication Data

Names: Scher, Rod, author.
Title: Ship of lost souls : the tragic wreck of the steamship Valencia / Rod Scher ; foreword by John E. Harrington ; illustrations by Molly Dumas.
Description: Essex, Connecticut : Lyons Press, 2024. | Includes bibliographical references and index.
Identifiers: LCCN 2024018229 (print) | LCCN 2024018230 (ebook) | ISBN 9781493081356 (cloth) | ISBN 9781493081363 (epub)
Subjects: LCSH: Valencia (Steamship) | Shipwrecks—British Columbia—Vancouver Island.
Classification: LCC G530.V35 S35 2024 (print) | LCC G530.V35 (ebook) | DDC 910.9164/33—dc23/eng/20240605
LC record available at https://lccn.loc.gov/2024018229
LC ebook record available at https://lccn.loc.gov/2024018230

∞™ The paper used in this publication meets the minimum requirements of American National Standard for Information Sciences—Permanence of Paper for Printed Library Materials, ANSI/NISO Z39.48-1992.

This book is dedicated to Frances Scher, who was both of her time and a very long way ahead of it. She raised a rebellious and troublesome boy as best she could by herself, and that worked out pretty well, most of the time.
Thank you for inspiring me to love literature, writing, and baseball.
And Johnny Mercer and Harry James and Duke Ellington and so many others.
More than a quarter of a century after your passing, I still wonder, "Would Mom be proud of this?" whenever I do just about anything.
I hope I did you proud this time around. I miss you.

Contents

Foreword

Rod Scher's Ship of Lost Souls: The Tragic Wreck of the Steamship Valencia reads like an adventure novel. If you're like me, once you start reading it you won't want to put it down. Unfortunately, the book isn't fiction; it is a true and well-documented recounting of the grounding and death of *Valencia* and the 170+ souls she carried. All were needlessly lost. Rod's story starts by introducing you to a small group of well-respected, experienced, hardworking, and caring people. Their decisions and actions, though well meant, are critical elements in the unfortunate chain of events leading to the tragedy. The book also introduces you to other "characters," including the SS *Valencia* herself, and the newfangled radio telecommunications technology that was in its infancy. Knowing these main players sets the stage for a fact-based story that is as relevant now as it was in 1906, because even in our modern world of immediate communication, social media, and precise electronic navigation, we still suffer from the same weak links that doomed the passengers and crew of *Valencia*.

Valencia's captain, crew, and passengers overlooked the lurking dangers of going to sea. The perilous ferocity of the sea is never to be trifled with. During my career in the US Coast Guard, I commanded USCG cutters and crews responsible for innumerable Search and Rescue operations. I sailed the Atlantic, the Gulf of Mexico, the Caribbean, and the Pacific. Then, in 2015, my wife, Janet, and I moved aboard our fifty-two-foot sailing yacht, *Tango*. For three years we refitted and equipped *Tango* for crossing oceans. Since then, we have crossed the Pacific twice, felt the power of the Gulf Stream along the US East Coast, and fought the northerly currents off California as we sailed south to Mexico. We crossed the always-turbulent Caribbean and piloted *Tango* through the wonder of modern engineering that is the Panama Canal. Through all these experiences I have learned two key things: First, while ships absolutely have personalities, it is the crew that makes or breaks a ship. The second thing is just how essential meticulous maintenance and training are. The crew must train and practice so they can instinctively react when bad things start to impact their day at sea. And bad things *will* eventually impact your day at sea, you can count on it. When that happens, you need to be trained so well that your reaction is both appropriate and instinctive.

US Coast Guard cutters, and *Tango* for that matter, are not particularly special as ships go. What is special is the maintenance and testing done to the ship's systems and the crew's training: firefighting, control of flooding, first aid, navigation, weather, use of rescue equipment, etc. I have come to believe that if I treat my ship right it will treat me right when the ocean is misbehaving.

Just after I reported aboard my first USCG ship, one of our main propulsion motors (it was a diesel-electric drive ship) caught fire as we were transiting the Chesapeake Bay. Now that was pretty scary. However, the crew knew exactly what to do. In a matter of a few minutes the fire was out; we never lost propulsion or maneuverability, and throughout the incident the many different teams communicated effectively. We were trained and prepared and had practiced for just such a scenario. What could have been a disaster became instead a lesson in discipline and training.

Several months later, while on patrol about two hundred miles off the East Coast of the United States, we responded to a mayday call from a USN ship that had caught fire. The crew could not control the fire and they abandoned ship because they didn't know how to use their equipment or how to fight the fire. We arrived on scene and our firefighting party boarded the Navy ship and quickly extinguished the blaze. The difference was that we knew how to use the firefighting systems and equipment. We also knew how to put the fire out and get things under control. We were experienced and well trained. Several hours after the fire was extinguished, with our help, the very shaken Navy crew was able to get underway and return to port. We sailed off, continuing our patrol.

During our first Pacific crossing aboard *Tango*, we suffered a knockdown when a squall line was able to sneak up behind us with—as far as we could tell—no warning. The helmsman was just starting to stand watch. Janet and I were cleaning up from dinner, when BOOM! *Tango* was on her side. The helmsman did what he thought was best, but the conditions were different from what he had experienced for the several weeks prior. I went on deck and immediately turned *Tango* into the wind so that she stood up out of the water. It took about forty-five minutes for the squall to pass. (You can read Janet's take on this incident on our blog: http://tinyurl.com/42x3j95p.) As it happens, our knockdown was completely avoidable, but that surprise squall could have ended with *Tango* losing her mast or sinking. Aboard *Tango* the helmsman is the lookout; this requires 360-degree situational awareness. Had our helmsman occasionally looked behind us, he would have seen the approaching weather. Before the squall hit, we could have reduced sail, then maintained a safe course until it passed. We only did one little thing wrong, but sometimes "one little thing" is all that stands between a sailor and disaster. However, we were lucky—or, better yet, we were *prepared*; good maintenance and my experience at sea broke a potentially catastrophic chain of events.

Radio and navigation equipment are *much* better today than in 1906. GPS, satellite data, and social media have all contributed to improved safety at sea. In 2023 the yacht *Raindancer* collided with a whale while sailing from Panama to French Polynesia. *Raindancer* sank in only fifteen minutes, but the crew did virtually everything right, despite the unpredictable nature of such an encounter. They knew what to do, where key things were, and how to communicate. *Raindancer*'s crew was prepared. They used a social media app (Boat Watch) and text messaging to report the accident. Their intelligent and timely use of today's technology over a satellite communication link enabled a successful rescue by a yacht only sixty miles away. *Raindancer*'s crew broke the chain of events that could have caused a catastrophe but in the end merely resulted in a frightening but relatively painless inconvenience.

Those who go to sea need to know that it is seldom the ship or the weather that causes a mishap. It is us. People are the weak link, and *Tango*'s knockdown is a great example. Alarmingly, avoidable shipwrecks occurred well after 1906 and continue to occur today. Consider *Titanic*, *Andrea Doria*, *Costa Concordia*, *Doña Paz*, and *Exxon Valdez*; all of these are similar to the SS *Valencia* wreck in that they were completely avoidable. The weather did not cause these wrecks, and neither did the vessels themselves. *People* caused them.

The single largest contributing factor in most marine accidents, by far the weakest link in the chain, is us humans; even when ships' systems function perfectly, *we* continue to fail. Poor crew training, placing schedule and efficiency over safety, failure to maintain ship systems (engines, steering, propulsion, communication, etc.), not being practiced with safety equipment (lifeboats, life rafts, firefighting equipment, life preservers), and permitting a lack of preparedness (not training crew and passengers regarding fire, loss of propulsion, lifeboat, and raft deployment, etc.) has repeatedly led to unnecessary death and damage.

Ship of Lost Souls details the characters and the chronology of the *Valencia* tragedy and summarizes the investigations and actions taken after the tragic loss of life. You'll find the book riveting as the tragedy unfolds. Like me you'll be disappointed in how slowly navigation systems and maritime regulations improved. For me, the book reawakened an awareness of my responsibilities as *Tango*'s skipper. Thus, it was not merely an enjoyable read, it was also a call to action.

Read this book. Enjoy it for the spellbinding nautical tale that it is, but be sure to take its lessons to heart.

CDR. John E. Harrington, USCG (Ret.)
Aboard S/V *Tango* off the coast of New Zealand
February 2024

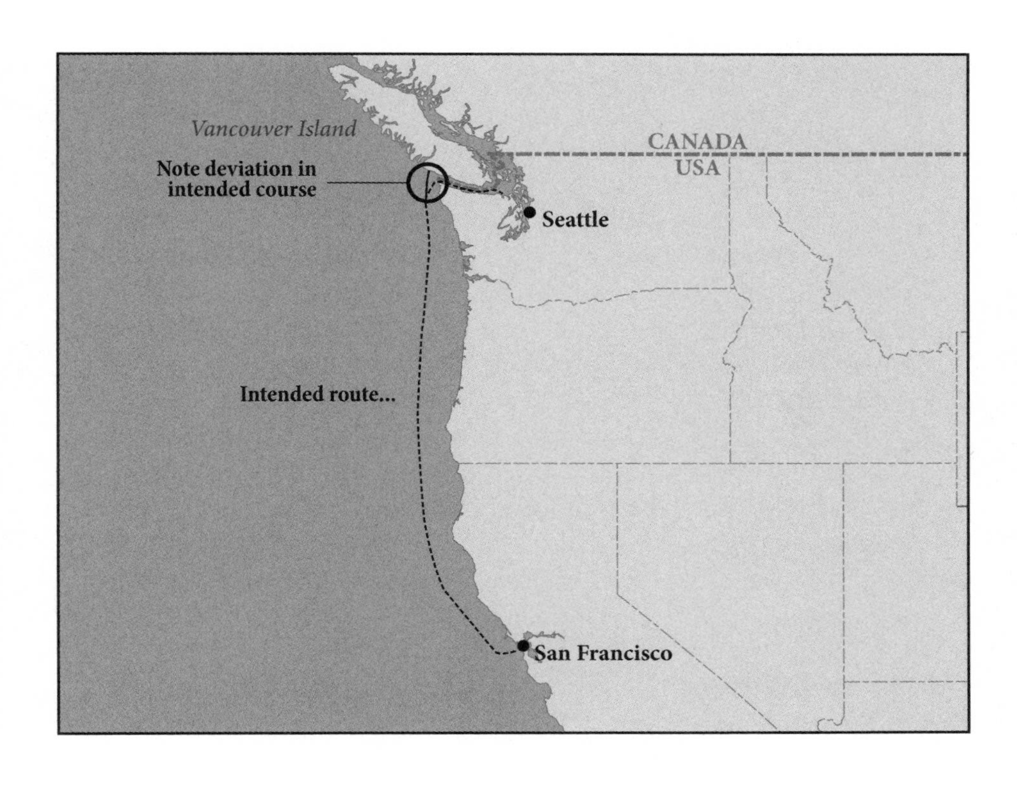
Vancouver Island
CANADA
USA
Note deviation in intended course
Seattle
Intended route...
San Francisco

Prologue

In some ways, this story—or at least our window on the story—actually begins twenty-six years *after* the sinking of the SS *Valencia*, during the summer of 1933. The wooded wilderness of Pachena Bay, British Columbia—near the place where *Valencia* struck the rocks that holed her iron hull—is calm, peaceful, and mostly empty. Few come here. For one thing, there is almost no way to *get* here, other than by boat. There is a narrow path up above the cliffs, part of what became known after the sinking of the *Valencia* as the Dominion Life-Saving Trail. A telegraph line runs in the trees along the path, but in 1933 it is down about as often as it is up, and few rely on it. The bay itself ripples and shimmers in the breeze; the tops of the trees, which grow almost to the rocky shoreline, sway gently. It is deserted, a quiet, tranquil place.

In fact, the whole world seems calm and largely at peace. In the United States, Prohibition has just ended and work has begun on San Francisco's Golden Gate Bridge. Franklin Delano Roosevelt has been inaugurated the thirty-second president of the United States. The Washington Senators will face the New York Giants in the World Series. (The Senators will lose, badly.) In the boroughs of New York, it is possible to walk through a neighborhood and never miss a moment of the action as announcers from WOR and WABC bring the games to life for their listeners, the static-ridden play-by-play drifting out of tenement windows opened against the oppressive summer heat. Bookish sorts might be reading John Steinbeck's *The Red Pony*, George Orwell's *Down and Out in Paris and London*, or Laura Ingalls Wilder's *Farmer Boy*. It is a seemingly gentle time, though it will not remain so for long.

Deep in the Canadian coastal wilderness stands the Pachena Point lighthouse. Its Fresnel lens blinks a blinding double-flash every 7.44 seconds as it guards a forbidding promontory fronted by one-hundred-foot cliffs. The lighthouse is in radio contact with other lighthouses along the coast, and two lifesaving stations stand not far away. The scene is desolate and wild, but the lighthouse, the radio antennas, and the lifesaving stations provide ample evidence that mariners stranded in that wilderness are not without recourse; the appropriate authorities have considered their plight and put into place infrastructure aimed at guarding their lives and effecting their rescue, should it become necessary. As desolate as the place seems, help is nearby.

A small boat bobs in the sparkling water, dipping and nodding in the wavelets. The vessel is empty, and it drifts and circles aimlessly in the breeze and in response to the hidden currents and eddies of the bay. The boat has not been seen for twenty-six years; it simply appeared in the bay one morning. Though its paint is chipped and worn away, on its bow it bears the still perfectly readable nameplate that proudly proclaims it *Valencia*'s No. 5 lifeboat.[1]

Where did it come from? Lifeboat No. 5 had been launched on January 23, 1906, from the steamship *Valencia* as she foundered on the rocks off the west coast of Vancouver Island, British Columbia, several miles from the bay where the boat reappeared all those years afterward. In the

This is the nameplate from the *Valencia* lifeboat found floating in Barkley Bay, twenty-six years after the sinking.

PHOTO BY LESLEY SCHER. USED WITH PERMISSION. COURTESY OF HEATHER FEENEY, MARITIME MUSEUM OF BRITISH COLUMBIA.

unfolding tragedy, most of the lifeboats—hastily lowered and ineptly manned—were destroyed almost as soon as they hit the water. Lifeboat No. 5, with boatswain Tim McCarthy in charge, had not been seen since it had been lowered into the water in a desperate and futile attempt to save the lives of people aboard the ship.

Where had the lifeboat been for twenty-six years? Some theorized that otherworldly forces were at work here; stories of ghost ships and skeletons found washed ashore made the rounds, and still do. Others, more realistically perhaps, theorized that the lifeboat had been washed into a cave where it had remained trapped until freed by a recent storm surge. No one could deny, though, that the reappearance of the battered lifeboat was . . . *peculiar*, and somehow oddly unsettling.

Neither the lifesaving stations nor the lighthouses nor the radio stations existed in 1906; if they had, the story of the *Valencia* and her 170 or so passengers and crew might have ended differently. As it was, the lack of such aids to navigation on both the Canadian and American coasts and the absence of lifesaving infrastructure together helped guarantee that 136 passengers and crewmen on the vessel—including every single woman and child on board—would perish in the wreck.

This is the story of that disaster, one made doubly tragic by the loss of women and children and then, somehow, made even more catastrophic by the fact that both the wreck itself and the ensuing lack of effective rescue efforts were the result of entirely avoidable errors. Mistakes—and perhaps shady deals—were made; after the fact, politicians and the press railed, and steps were taken to avoid a future occurrence of this sort. In the end, though, that didn't matter to the crew or passengers, because almost all of them were dead. Many drowned or died of hypothermia, some were crushed by massive spars and beams as the ship broke up in the waves, some were pounded against the rocks until their bodies were limp and bloody—and all within yards of land and within sight of would-be rescuers. Why? This book attempts to explain the inexplicable: How did so many people, entrusted to the care of the finest ship's officers and taking advantage of the best technology that was available at the time, end up dying? Why did women and children, lashed to the rigging, perish while watching their putative rescuers sail away? Why were men battered to death on the rocks only feet from the shore? Most of all, why did no one help?

PART I

A Storm in the Distance

Tragedy is restful: and the reason is that hope, that foul, deceitful thing, has no part in it.

Jean Anouilh, *The Play as Theater*

Chapter 1

The Professor

Frank Forest Bunker was born in 1873, in the tiny village of El Dara, Illinois (even in 2020, its population was only about ninety), to Theodore Clairmont Bunker, a small-scale farmer, and Clara Wood. The middle child of five siblings, Frank distinguished himself quite early as an exceptional thinker and a remarkable student; in a family of hardworking parents and relatively uneducated brothers and sisters, he was destined for academia, and he knew it early on.

Things would not have been easy for a family of seven in those years. When Frank was born, a depression gripped the United States, and it did not ease until 1879. During the depression of 1873, which was largely brought on by rampant speculation in railroad construction, the world-wide economy failed as, among other things, construction expenses exceeded and outpaced financing. Until the market crash of 1929, it was was the worst economic failure to strike the world, and families such as the Bunkers did all they could to scrape by. Luckily, by the time the young man entered his teenage years, the country had mostly recovered, and there was a little money to send him to college, and plenty of jobs available to help him earn his way.

Interestingly, the same year that Frank Bunker was born in Illinois, Lee de Forest was born in Council Bluffs, Iowa, only about 350 miles away. De Forest would become a radio innovator of some note, eventually inventing the vacuum tube, which would revolutionize broadcast radio—but too late to help the passengers and crew of the *Valencia*.

Inventor and Marconi competitor Lee de Forest
ILLUSTRATION BY MOLLY DUMAS. USED WITH PERMISSION.

A year *after* Bunker's birth, the man largely responsible for the eventual success of commercial radio, Guglielmo Marconi, was born in Bologna, Italy. The two men, de Forest and Marconi, would lead lives that would intersect as they became both partners and, ultimately, rivals in the nascent world of radio communications. If they had succeeded in their endeavors just a bit earlier, more ships might have carried radio equipment in the early 1900s, and the vessel on which Frank Bunker sailed in 1906 might not have struck the rocks off of Vancouver Island.

As expected, Bunker became a teacher and, in 1899, when both of them were twenty-six years old, the young man married Isabelle Ball in Tulare, California, then a modest settlement of some 2,200. The town was itself only twenty-seven years old, having been founded in 1872 by the Southern Pacific Railroad. In fact, the railroad—and its desire for a transportation hub—was the only reason for the town's existence in the first place. The town had existed only one year longer than Bunker himself.

Bunker had been working in Tulare, California, as a teacher but had always aspired to a career as an administrator. In 1900, he and his new wife moved to Santa Rosa, where Bunker moved up the educational hierarchy, becoming the vice principal of Santa Rosa High School, one of the oldest high schools in California, having originally opened in 1874.

But Bunker's aspirations were lofty, and he was not satisfied being a lowly vice principal. His career goals—and his ego—demanded more than that. He envisioned being a principal, perhaps a school superintendent, a researcher, maybe even a celebrated author. The boy from the little village in Illinois was out to make a name for himself.

Santa Rosa High School in the mid-2000s. The building still retains a bit of its 1870s-era flair.

In time, all of these dreams would be realized, and along the way, he would create a family and continue to climb the career ladder. He and Isabelle would have two children, Dorothy in 1901 and Frank Jr. in 1904. In the early 1900s, the Bunker family would move to San Francisco, where Frank Sr. would become the principal of the recently opened La Conte Grammar School near downtown, just a stone's throw from UC Berkeley and today part of the Berkeley Unified School District. One of the reasons we know that Bunker was principal of the school in 1902 is because that year the Santa Rosa *Press Democrat* recounted a minor brouhaha that resulted after Bunker was reported to have "thrashed" a La Conte Grammar School student. His friends—academic and otherwise—defended him, with the paper noting: "Bunker's many friends here are loth [*sic*] to believe that he was unduly harsh or that the punishment administered was not merited."[1] In any case, the resulting notoriety does not seem to have hindered his steady rise through academia.

Bunker's career arc appears to have been inevitable by that point. In 1904, he would accept a position as a professor at the State Normal School at San Francisco. A "normal school" is what would today be called a teachers' college. An odd relic of those times can be found in older street names: towns that once had a "normal school"—and that's most of the ones that were large enough to support a small college—often retain a Normal Street or Normal Boulevard or some such thoroughfare as a holdover, even though the school itself may have closed down or changed its name.

Bunker was now not a mere teacher but a professor—a teacher of teachers. He would forever remain a professor and administrator; his days as a mere grammar school teacher were now behind him.

In 1906, Bunker again moved up in status, accepting an offer to become the assistant superintendent of schools in Seattle, Washington. In January of that year, he and his young family embarked on what was meant to be an enjoyable and brief journey from San Francisco to Seattle so that he could take up his new post. It would be an invigorating three days at sea—long enough to give the Bunkers the feel of a true sea voyage but not so long that boredom could set in and not so far out at sea that danger could threaten.

But danger did threaten. The ship on which he and his family sailed would never make it to Seattle. En route, Bunker would lose his wife and his two children, his family dying in front of his eyes, drowned in the storm-tossed waves or succumbing to hypothermia while attempting to flee the SS *Valencia* as it broke up on the rocks off the west coast of Vancouver Island, British Columbia. Bunker would survive, and his actions after he made it to shore would forever brand him either a hero or a coward, depending on who was doing the telling. The truth might be found, as many truths are, in some gray middle ground between the two. He may have been neither or some of both. Or perhaps he was a hero, but an unlikable one; we never quite know what to do with our unlikable heroes. Of course, the complete truth is shrouded in mystery, political machinations, and finger pointing, but this book will attempt to describe Bunker's role in the tragedy as objectively as possible.

Chapter 2

Solid, but Not Beautiful

Many ships are beautiful; shipbuilding, after all, is a breath-takingly complex undertaking, part science and part art, in which form and function often combine to create, through some sort of magical alchemy, vessels as stunning as they are practical. Think, for example, of a regal clipper, a "greyhound of the sea," with her sharp prow and narrow beam, tall wooden masts festooned with dozens of dazzling white, full-bellied sails billowing in a stiff wind. Or a majestic ocean liner, with her grand and majestic bulk, her powerful engines driving her effortlessly through the churning seas, a symbol of stately elegance. Or even a small but refined pocket yacht, her graceful and flowing lines a testament to some distinguished designer, perhaps a modern master like John Alden or Ted Brewer or Nate Herreshoff. Some vessels, perhaps most of them, are impressive either by virtue of their remarkable beauty, their breath-taking performance, or their monumental grandeur, or, in rare cases, all three. This was not the case with the SS *Valencia*.

When *Valencia* rolled off the ways at Philadelphia's Cramp and Sons in 1882, she was solidly representative of the well-built vessels that the company had been producing since it was founded in 1830 by twenty-three-year-old William Cramp. Cramp, born in Kensington, Pennsylvania, would go on to become the preeminent builder of iron-hulled vessels in the nineteenth century. By the time the firm launched *Valencia*, Cramp and Sons had built and launched roughly one hundred ships, including pleasure boats, passenger liners, and warships. Ten or so of those vessels had been built for various US armed services, the majority of those during the Civil War.[1]

Valencia may have been solid, but she was most assuredly not beautiful. Her narrow beam and overly long bow made her look spindly and ungainly, and indeed she was not the most stable of vessels; her half-decks and her boxy wheelhouse contributed to her off-center, unbalanced look,

Shipbuilder William Cramp about 1870, in his sixties

and the lone steam funnel jutting out of her deck amidships simply underscored the fact that, as new as she was, she was underpowered from the very beginning. She could only be called a "greyhound of the sea" if one had in mind a dog that had been born a runt and then starved and mistreated—an ailing canine representative of a mangy, deteriorated line of seagoing animals that would certainly not win any races and that might not even *survive* a race.

Like her sister ships, *Valencia* also lacked the luxury and amenities found on the larger oceangoing passenger liners: the gymnasium, the library, the many gaming parlors. She was, compared to those seagoing titans, stark and utilitarian, a second-class conveyance in all respects.

These days, just as when a home, apartment, or commercial building is constructed, ships are built according to strict reliance on construction codes applicable within the country of construction and also rules and standards agreed upon by shipbuilders, lawmakers, and inspectors around the world.[2] But in 1882, when *Valencia* was built, there were few agreed-upon codes that shipwrights were required to follow. There were standards, practices, and construction techniques long followed by professional shipwrights, but they had not been codified into law.[3] These days, of course, there are multiple sets of codes and rules, such as the ones promulgated by treaties such as SOLAS (The International Convention for the Safety of Life at Sea), which cover how and where things such as bulkheads must be built, and they are very specific. For example, one part of one rule says:

> There must be no doors, manholes, access hatches, ventilation ducts or any openings on the collision bulkhead below the bulkhead deck. However, the bulkhead can be allowed to have only one piercing below the bulkhead deck for the passage of one pipe to cater to the fluid flow to the forepeak ballast tank. The passage of the pipe must be flanged and must be fitted with a screw-down valve which can be remotely operated from above the bulkhead deck. This valve is usually located forward of the collision bulkhead. However, the classification society certifying the ship may authorise a valve aft of the bulkhead provided it is easily serviceable at any condition, and is not located in the cargo area.[4]

Remember that this is one part of one rule and that each rule might consist of several such sets of instructions. If you ask plumbing or electrical contractors today about the myriad codes and requirements that they must follow when building a home, they will understand exactly how strict the requirements of building a seagoing vessel can be. As with building a house, building a ship is a complicated—and sometimes frustrating—enterprise.

All of which is to say that the standards of the 1900s, as rudimentary as they might seem to us, were largely absent when *Valencia* took shape on the ways at Cramp and Sons. She was a well-built vessel for her time, but she was lacking in safety features that would become common only a few years later.

Primary among *Valencia*'s lack of safety features was the absence of multiple waterproof bulkheads that would, in more modern vessels, be used to ensure that a gash in the hull could flood only a small number of the ship's compartments. *Valencia* possessed only four such bulkheads, according to the commission's report—two protected her engine room and boilers—while the location of the others was not noted.[5] That meant

A view of the port side of the steamship *Valencia*
ILLUSTRATION BY MOLLY DUMAS. USED WITH PERMISSION.

that the rest of the ship—much of its entire 252-foot length—would flood if the hull were pierced somewhere other than those areas.[6]

In addition, the commission investigating the incident questioned not merely the number but the *condition* of *Valencia*'s bulkheads. It had been charged that the bulkheads she did possess were not structurally sound. Oddly enough, this may have been partly due to the fact that *Valencia* had, during her long career as a cargo vessel, transported large quantities of *sugar* in her holds. It was the opinion of chemist Malcolm Faziell, who worked for Vancouver's British Columbia Sugar Refining Company, that if raw sugar were transported in the ship, and if it contained impurities, the acids in those impurities—especially if the sugar had gotten wet—could have had a destructive effect on the bulkhead, resulting in a "honeycombing" effect, weakening the steel bulkheads. The argument *against* this theory is that the process as described would have generated a terrific—and terrifically unpleasant—odor, but there is no record of anyone having complained about such an odor. Nonetheless, after striking the rocks, the ship took on water very quickly, and her hull was speedily destroyed by the wind and waves. The ship's rapid disintegration seemed suspicious to some and led to questions about the integrity of her hull and bulkheads.[7]

Nonetheless, investigators noted that *Valencia* had been inspected multiple times, the most recent inspection having taken place only three weeks before her last voyage. The inspectors determined that her equipment was generally in excellent condition, including her engines, machinery, and hull, which had recently been overhauled. There was no mention of any work being done, or needing to be done, on her bulkheads.[8]

Still, the January 30, 1906, issue of the *Seattle Star* questioned the integrity of the *Valencia*'s bulkheads, quoting unnamed "local steamboat men" as saying, "Had they [that is, the bulkheads] been of proper strength, the *Valencia* would not have become completely submerged." The article pointed out that the law required steamers carrying passengers to have at least three watertight bulkheads and that those bulkheads must reach to the main deck or, in a vessel with multiple decks, such as *Valencia*, "to the deck below the main deck." The paper then quotes further unnamed sources as saying that "the bulkheads were not as strong as the law requires, and when the water rushed in, they broke, permitting water to fill the entire hull of the vessel."[9] Of course, the law may not have applied to vessels built twenty-four years prior to *Valencia*'s 1906 voyage, and multiple watertight compartments such as those described may not have sufficed, in any case.

Valencia was 252 feet long, iron hulled, with a beam of 34 feet and an unloaded draft of 19 feet, and she was licensed to carry 286 passengers. She had three cargo holds, and watertight compartments protected her engine and boiler room; but, significantly, there were no such compartments elsewhere. *Valencia* was not fitted with a double bottom, and her bulkheads, as noted, were alleged by some to be somewhat insubstantial.[10] She was a single-screw vessel, which limited her maneuverability, especially in tight quarters. For the sake of comparison, the SS *Edam*, launched by Nederlandsche Stoomboot—the company now known as Holland America—in the same year, measured 328 feet in length and had a beam of over 41 feet.[11] The SS *Normandie*, later known as *La Normandie*, built by Barrow Shipbuilding in 1882, was a whopping 459 feet long and boasted a beam of 49 feet.[12] Both larger liners were equipped with twin screws and multiple engines.

Valencia was small, slow, and dowdy, but as inelegant as she might have been, she was not unfit for the job for which she had been designed: the warm-water run between Venezuela and New York City. Owned at the time by the Red D Line, as was her sister ship, *Caracas*, *Valencia* plied the coastal waters down into South America and back to the United States for several years. The Red D Line had been operating service to Venezuela since 1839, and the addition of more modern steamships, including *Valencia* and *Caracas*, allowed the line to carry not only passengers but also mail and cargo, often on the same vessel. In fact, when *Valencia* struck the rocks off of Vancouver Island, British Columbia, in 1906, many of the passengers and crew found it difficult to maneuver on the decks of the ship partly because a cargo of cabbages had broken free and were now rolling about the decks and getting underfoot.[13]

In the end, *Valencia* was built as well as most other vessels—and surely better than some—but she was showing her age. Over time, iron plates rust. Steel rivets weaken and crack. Bulkheads that may have sufficed

Valencia's sister ship, SS *Caracas*, ran aground in Yaquina Bay, ironically, shortly after being renamed the SS *Yaquina Bay*.

at first may not be up to the task twenty-four years later, and especially not when subjected to the turbulent waters and ferocious weather of the fierce Pacific Northwest coast—which is not the environment for which she had been designed, in any case.

The Venezuela run was an efficient and cost-effective use of the Red D fleet, and *Valencia* became an integral part of that fleet, making her maiden voyage in May 1882, sailing from New York City to Caracas via Laguayra and Puerto Cabello, Venezuela.[14] Twice a month, the two sister ships would undertake journeys of roughly twenty-six days, ferrying passengers, mail, and cargo back and forth between Venezuela and New York. In those waters, and staying close to shore, *Valencia* and her somewhat smaller sister vessel served the Red D Line well, but things were about to change, and *Valencia*'s history—and her reputation—would turn spotty as the years went on.

Chapter 3

"This Nice Achievement"

Edouard Branly was a cranky old man—and, to be truthful, apparently a bit touchy even as a *younger* man. Then again, the French physicist, physician, and inventor was probably entitled to a certain measure of irritability: indisputably brilliant, he had little patience for those of middling intellect, and he was determined to be appreciated and, perhaps even more importantly, to be paid what he thought he was worth. Perhaps this explains why, in the middle of a successful but poorly compensated career as a physicist, he dropped everything and began attending medical school, eventually earning a medical degree. Medical doctors, he correctly reasoned, were well paid and were usually provided excellent working facilities. Branly eventually went back to his physics experiments, and it's a good thing for the future of radio that he did. We don't hear much about Branly these days. When we think of early radio experimenters, we think of men such as Tesla, Edison, and—perhaps most of all—the brilliant and spectacularly shrewd Guglielmo Marconi.

But radio owes more to Branly than most of us realize. In the mid- to late 1880s, while *Valencia* was engaged in her voyages between New York and Venezuela (and then later carrying troops to the Philippines during the Spanish-American War; see chapter 4), and while a teenaged Frank Bunker was preparing for a career as a teacher, Branly was experimenting in his poorly equipped—or so he complained—laboratory in the Sorbonne and then in a newer lab at the Catholic University in Paris. One assumes that the latter lab was better equipped and furnished more to Branly's liking, and that his salary was higher than what had been offered by his previous employer. Then again, according to some sources, there may have been yet another reason for Branly to change employers: at the Sorbonne, he was apparently under some pressure to marry his boss's daughter, a union about which Branly was seemingly less than enthusiastic.[1]

Among many other devices and ideas Branly developed (he published more than one hundred papers on a wide variety of topics) was an odd-looking gadget that was destined to change the world of communications: the "coherer." This was essentially just a glass tube filled with iron filings and connected in a circuit with a battery. When an electromagnetic current was detected nearby—even in another room, as it very significantly turned out—the filings would "cohere," that is, they would clump together, thus forming a conductive path. This new path would allow a current to flow through the coherer, and that current could then be detected by a galvanometer, a device that measured small amounts of current and that would eventually evolve into the twentieth-century ammeter. Note that Branly did not call the device a "coherer," fellow radio experimenter Oliver Lodge did; the French scientist referred to it as a "radio-conductor," a possibly more accurate and certainly more mellifluous name. Nonetheless, the term "coherer" caught on and was used quite extensively for a number of years. Regardless of what it was called, it's easy enough to see that, connected to a speaker of some sort, such a device could react to the presence of an electrical wave by sounding a Morse code "click" or, when connected to an appropriate mechanical device, could punch a code onto a paper tape. (Unfortunately, after each "click," the particles would have to be "de-cohered," so early radio setups included an electromechanical device—invented a few years later by Lodge—that literally tapped on the tube to reset the iron filings.) Branly had just invented the mechanism that would make radiotelegraphy possible. On Branly's shoulders, and on the shoulders of this invention, would stand such radio pioneers as Oliver Lodge, Thomas Edison, and Guglielmo Marconi.

As Newton noted, scientists really do stand on the shoulders of those who came before them. Branly, though surely as inventive as they come, was essentially repeating and expanding upon experiments made some

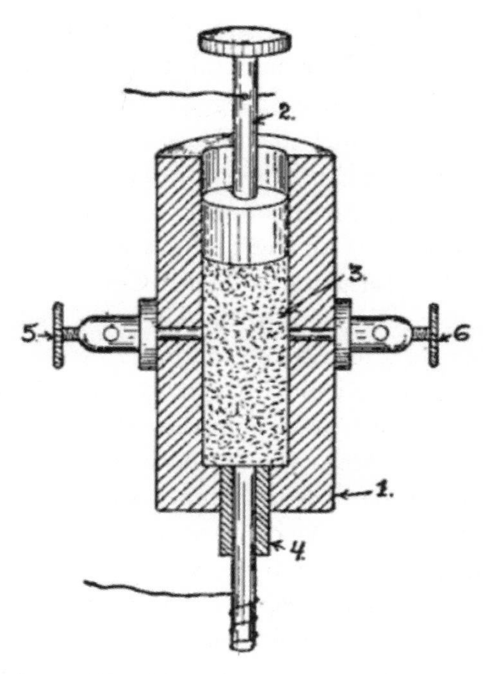

An 1890 diagram of one of Branly's early coherers

years earlier by Temistocle Calzecchi Onesti, an Italian physicist and inventor who showed that iron filings contained in an insulating tube will conduct an electric current under the action of an electromagnetic wave. And Lodge, acting after—and occasionally with the help of—Branly, continued Branly's experiments and eventually invented the antenna coil or inductance variable, introducing the concept of tuning in order to select a desired frequency. Branly would eventually sell the patent to this device to Marconi, his sometime partner and occasional competitor.

Of course, the king of early radio was Guglielmo Marconi. An avid and relentless inventor and tinkerer, Marconi would eventually hold some thirty or more patents, most dealing with the transmission and reception of radio waves. He is often credited as "the inventor of radio," but, as we have seen, that's a bit of an oversimplification. Not only did he owe much to those who came before him, he was not unaware of that fact: in 1899, when he made the first radio transmission across the English Channel, a distance of about twenty-one miles, he explicitly credited Branly in his transmitted message: "Mr. Marconi sends to Mr. Branly his regards over the Channel through the wireless telegraph, this nice achievement being partly the result of Mr. Branly's remarkable work."[2]

What Marconi lacked in originality, he more than made up for with an incredible perseverance and dedication to hard work, which he combined with astounding and wildly prescient business acumen. He founded the Wireless Telegraph & Signal Company, which eventually became the Marconi Company, in the United Kingdom in 1897 and began looking for ways to increase the range of his radio signals and to convince officials in the United Kingdom to install his radio stations on land and in the country's merchant vessels and warships. At first, he concentrated largely on ocean vessels and marine transmissions simply because he had discovered that in Europe, government-operated postal services had a monopoly on the delivery of messages of all sorts, so initially he was able to transmit radio waves great distances *only* over water.[3]

The state of the radiotelegraphic art progressed quickly, and 1899, only two years after Marconi founded his company, turned out to be a banner year for its overall development. In that year, British warships using Marconi gear exchanged messages at an astonishing distance of seventy-five miles. At roughly the same time, a wireless radio experiment was conducted by the US Life-Saving Service on board *Lightship 70* stationed off San Francisco, California, and that November, Marconi, on board the SS *St. Paul*, radioed a shore station from fifty miles away as *St. Paul* made her way to the English coast. The station received her message, making *St. Paul* the first ship to have her arrival radioed to shore by wireless.[4]

About the time that Marconi founded his company, *Valencia* was chartered by the Ward Line to take the place of the Red D liner *Niagara*. *Valencia* was then "attacked" off Guantánamo Bay by the Spanish cruiser *Reina Mercedes*, which—in the midst of tensions that would soon lead to the Spanish-American War—fired two warning shots at her.[5] Immediately, the American flag was raised on *Valencia*'s stern, and it was later determined that the shots were fired by the Spanish vessel specifically in order to intimidate *Valencia* into raising her colors, as was required by maritime law during wartime.

Guglielmo Marconi pictured in 1909, at age thirty-five

Valencia should have been showing her colors to begin with, of course, especially at a time when tensions were running high. Vessels that fail to display their colors may be subject to attack by enemy forces, because those enemy forces cannot be sure of the nationality of a vessel that is not

The SS *St. Paul* under way near New York harbor in 1895

The Spanish cruiser *Reina Mercedes*, pictured here sometime before 1898, was launched in 1887 from the naval shipyard in Cartagena, Spain.

flying her colors; when a vessel flies her colors, she identifies herself and indicates that she is not a threat. The *Reina Mercedes* obviously knew that *Valencia* was a civilian ship, which is why the Spanish vessel fired only warning shots and ceased fire once the American vessel raised her colors.

Nonetheless, this was sloppy handling on the part of the American ship's captain, and he should have known better. During the 1897 incident, the captain was *not* Oscar M. Johnson, who would be the ship's master when she ran into trouble several times in later years, but the mistake was a sign of things to come.

Chapter 4
Valencia Goes to War

The Spanish-American War, of which the *Reina Mercedes*'s mock "attack" on *Valencia* was but one harbinger, began on April 21, 1898, after the sinking of the USS *Maine* a few months earlier in Havana Harbor, and ended on August 13, 1898, with the Treaty of Paris.[1] It was caused, as are so many wars, by a complex mélange of issues and motives, primary among which were the United States's need for coal, a desire for overseas expansion (itself perhaps an expression of what had come to be known as Manifest Destiny), and passions inflamed by an ongoing revolution in Cuba—and what many in the United States saw as the government's mistreatment of citizens there during Cuba's revolt against Spain.[2]

The war was well received by many Americans, with populist ire fanned by sensationalist newspapers owned by William Randolph Hearst, Joseph Pulitzer, and others. Hearst, in fact, was reputed to have cabled illustrator Frederic Remington, "You supply the pictures, and I'll supply the war," suggesting that at the time—and some would say perhaps still—the media often favored dramatic, and perhaps sensationalized, stories over objective reporting. (The war offered, among other things, a chance for the United States to show off its military might; John Hay, the US ambassador to Great Britain, described it as "a splendid little war!" Roughly two thousand or so Americans died during the "splendid little war," most of them laid low by illness rather than by enemy fire.)[3]

These days, the US Army maintains a fleet of ships larger than that of most navies, but in the nineteenth century, and even as recently as World War I and World War II, the Army had to charter a large number of commercial vessels in order to accomplish the transport of troops and supplies to various locations around the globe.[4] During the Spanish-American War, the US Army's Quartermaster Department chartered some ninety-five different vessels to serve in various theaters of the war. Most were troop transports—many of these were passenger liners sporting hastily added machine guns or other armaments—but that number also included hospital ships and other utility vessels that had been hired to serve in the conflict.[5]

Valencia was chartered by the US Army Transport Service on June 19, 1898, which paid the owners $650 per day (about $25,000 today) for her use, mainly as a troop transport. She was used to carry men and supplies from San Francisco to the Philippines, but she also occasionally carried troops from Cuba to Savannah, Georgia, all under the auspices of what was known as the Third Philippine Expedition.

At 282 feet, *Valencia* was smaller than many of the other chartered vessels, but she was capable of transporting some six hundred troops and a couple dozen officers to the war zone, as well as their equipment and supplies.

Cramp and Sons, *Valencia*'s builders, were well represented during the war. In addition to *Valencia*, some of her sister ships were also chartered by the Army. *City of Puebla*, for instance, was chartered on June 23, 1898, at a cost of $900 per day. Some confusion arises because Cramp and Sons's official history has occasionally been read (or misread) to imply that the company took credit for building the USS *Maine* itself.[6] The misunderstanding may result from the fact that Cramp and Sons did in fact build a battleship named *Maine*; designed in 1898, construction began in the fall of 1899 (*after* the Spanish-American War had ended), and resulted, eventually, in the production of a new 388-foot battleship of 12,500 tons. Again, keep in mind that this ship, designated BB-10, was built after the original *Maine* had been sunk. The newer vessel served for a time as the flagship of the Atlantic Fleet and cruised with the Great White Fleet as far as San Francisco. Used as a training ship during World War I, this latter USS *Maine* was decommissioned in 1920.

Valencia served admirably during the brief war; there are no known reports of problems with the ship, her officers, or her crew, but overall,

The SS *Valencia* loaded with troops in 1898 as she departs San Francisco

the exercise itself was rife with disasters, both realized and potential.[7] The chartered vessels, once in Cuba or in the Philippines, sat in the hot sun for weeks with inadequate sewage facilities: keep in mind that dozens of horses were often on board each vessel in addition to the men and that few provisions were made for the treatment and disposal of wastes, either human or animal.[8] Further problems became apparent when it was time to land the troops carried on the vessels. Not enough smaller boats were available to carry the men to shore; this made offloading the ships a time-consuming and dangerous undertaking. In addition, the lack of

Valencia's sister ship, *City of Puebla*, under way in Puget Sound circa 1900

smaller vessels, including lifeboats, meant that men aboard a torpedoed or otherwise damaged vessel would be unable to save themselves. Luckily, torpedoed vessels were a rarity.

In the end, the US government spent over $7 million (about $258 million today) to transport about ninety-two thousand men from one point to another—some of the men having been transported multiple times. *Valencia* acquitted herself well, carrying the men, mounts, and equipment of the 1st North Dakota Volunteers, the 1st Washington Volunteers, and the 2nd Illinois Volunteer Infantry safely to the Philippines or from Havana back home to Savannah, Georgia.[9]

Chapter 5

"You Have to Go Out, but You Don't Have to Come Back"

When *Valencia* was sailing her New York–Venezuela route in the 1880s and 1890s, and while young Frank Bunker was pursuing a course of study that would lead to his teaching degree, lifesaving infrastructure in both the United States and Canada was still embryonic, with new equipment and approaches evolving regularly but slowly. As with radio, another technology that could have aided the *Valencia* passengers and crew had it been available, lifesaving tools and infrastructure were at the time rudimentary and not nearly as helpful and efficient as they would one day become. Most importantly—and most unfortunately, as far as *Valencia* was concerned—lights and lifesaving stations were few and far between, especially in the Pacific.

Nonetheless, improvements were slowly being made to enhance the safety of mariners, even as early as the 1880s. In 1884, for instance, the Bureau of Navigation formed under the Department of the Treasury. This agency helped build and standardize lighthouses, and by 1885, the United States boasted 1,248 major lights, 1,745 minor lights, and an estimated 5,000 buoys in service as aids to navigation in American waters.[1] In 1889, the first incandescent light used in American lighthouses was installed at the Sandy Hook, New Jersey, lighthouse.[2] New, better lighthouses, such as the one at Sandy Hook, were being built rapidly, and the technologies used in those lighthouses were evolving almost yearly.

At the same time, while neither the US nor Canadian Coast Guards yet existed, there was nonetheless a patchwork of professional and semi-professional lifesaving services in both countries, staffed by dedicated and skilled crews who repeatedly risked their lives launching lifesaving craft from storm-battered beaches in defiance of the odds against them. Time after time, the exhausted men (at the time, they were all men) of the US Life-Saving Service—then another division of the Department of Treasury—dragged their heavy boats, placed on wheeled carriers that were almost as heavy as the boats themselves, through the sand and then into the surf, and set out to rescue stranded sailors and passengers. And they *were* helping: In August 1899, a lifesaving "surfman" with the unlikely name of Erasmus S. Midgett, of the Gull Shoal Life-Saving Station in North Carolina, single-handedly rescued ten people from the grounded bark *Priscilla*. Midgett was awarded the Gold Life-Saving Medal for his actions. In September of that same year, lifesaving crews at three stations in Delaware assisted twenty-two vessels, saving 194 persons without the loss of a single life.[3]

Things did not always go well for the US Life-Saving Service, especially in the early days. In January 1878, eighty-five persons died in the wreck of the *Metropolis*, off the coast of North Carolina. The ship's captain denounced the (apparently lackadaisical) efforts of the US Life-Saving Service, requesting that they be censured "in the severest terms," and the New York *Daily Tribune* accused the service of "scandalous inefficiency."[4] From the Life-Saving Service's public relations perspective, it did not help matters that, while the so-called professionals flailed about ineffectually, an enthusiastic Newfoundland dog plunged into the surf and dragged a half-drowned man to shore.

The Life-Saving Service improved over time, but the odds were often against them. Crewmembers regularly drowned or died of hypothermia, partly because the service accepted few excuses. It didn't matter how stormy the seas, how large the waves, or how bad the storm: the men of the US Life-Saving service were expected to launch their surfboats regardless of the conditions. This gave rise to the service's grimly humorous—but also chillingly accurate—unofficial motto: "You have to go out, but you don't have to come back."[5]

Surfman Erasmus Midgett sits on the deck of the wreckage of the grounded *Priscilla*.

The first surfmen's deaths recorded since 1870, when the service began using paid crew instead of volunteers, had occurred back in March 1876: in that year, seven surfmen perished in an attempt to rescue the crew of the *Nuova Ottavia*, an Italian vessel grounded near the Jones Hill North Carolina Life-Saving Station. This incident also marked the occasion of the death of Jeremiah Munden, the first African American surfman to die in the line of duty.[6] The *Nuova Ottavia*, meanwhile, mysteriously capsized in front of horrified onlookers who watched helplessly from near Currituck Beach, North Carolina, as the tragedy unfolded.[7]

Risk aside, the Life-Saving Service was improving, but there was a serious logistical issue looming, and it was one that would have dire results for the SS *Valencia*, its crew, and its passengers: By far the largest of the US Life-Saving Service's thirteen regions was the Pacific district. Covering all of California, Oregon, Washington, and Alaska, it

This image, from a 1906 postcard, depicts the launch of a shore-based lifeboat from a station in Chatham, Massachusetts.

was, inexplicably, served by a fairly small complement of only nineteen stations. This compares to the much smaller district made up of Maine and New Hampshire, which was served by fifteen stations, and the Long Island district, which was home to thirty stations.[8]

The Pacific Ocean is not merely the world's largest ocean, it is in fact the largest single geographic feature on Earth, period. It is, as one writer points out, "larger than the Atlantic Ocean, Indian Ocean, and Great Lakes combined."[9] The Pacific Ocean is so large, and its biomass so complex and comprehensive, that Herman Melville called it "the tide-beating heart of earth."[10] And in spite of its name, it is often stormy and wild, especially in the area of the US Pacific Northwest. While West Coast storms are generally less frequent than on the Atlantic Coast, Pacific storms tend to be wilder and to last longer. Wind-generated waves in Northern California have been known to tower up to forty feet and more in height, and the waters off the coast of Northern California, Oregon, and Washington State have been the scene of some of history's worst—and most tragic—wrecks.[11] It is not for nothing that the waters stretching northward from Tillamook Bay, Oregon, to the mouth of the Columbia River have become known as "the graveyard of the Pacific."

One would think, given the enormity of the area to be patrolled and the ferocity and frequency of the storms along the coasts of Oregon and Washington, that lifesaving stations would surely be placed at strategic

Another victim of the "graveyard of the Pacific," the remains of the *Peter Iredale,* a four-masted steel bark that went down in October 1906, sit half-buried on the beach near Oregon's Fort Stevens State Park in 2012.
COURTESY OF CHARLES KNOWLES, OF MERIDIAN IDAHO, USA, LICENSED UNDER THE CREATIVE COMMONS ATTRIBUTION 2.0 GENERIC LICENSE

US locations, perhaps just east of *Valencia*'s route, say, near Cape Flattery, Washington, near the mouth of the Strait of Juan de Fuca. One might further assume that, on the Canadian side, on the coast of Vancouver Island, British Columbia, a corresponding series of stations would be built and manned by Canadian lifesavers. That did not occur—or at least it did not occur in time to save the men and women of the *Valencia*. Just as there was no radio communication when *Valencia* struck the rocks off of Vancouver Island, there were also no lifesaving stations and only one lighthouse near enough to do her any good. She sailed alone, and she and her passengers and crew died alone.

Chapter 6

Radio Comes of Age—but Too Late

The dawn of the century heralded several developments, some of which might have affected the fate of the SS *Valencia*. Of course, many new technologies and changes appeared that would have had little or no effect on the ship's fate: in 1900, the US population exceeded seventy-five million for the first time, and the burgeoning population, or perhaps it was the transition to a new century, seemed to set off a landslide of innovation in the country.

First, Kodak introduced the Brownie camera, triggering a revolution in the commercial popularity of photography. Suddenly, *everyone* could—and did—take photographs. Then, in 1902, Willis Carrier invented air-conditioning. In that same year, Arthur Pitney created the postage meter, and the windshield wiper was invented and patented by Mary Anderson, an Alabama woman who didn't even drive.[1]

It was a busy and bustling time. In terms of technology, it was shaping up to be a blockbuster of a decade: the 1900s saw the invention of the airplane, the radio (more about that momentarily), the telephone, the toaster, and the washing machine. Even the disposable razor and the vacuum cleaner made their debuts in the early 1900s. It was as if a new industrial revolution were underway.

As far as technology related to sailing and to maritime safety—that is, in terms of mechanisms and developments that might have played a part in the survival of the *Valencia*'s passengers and crew—the new century brought similar new and important developments; unfortunately, most of them came too late.

Milton J. Burns's painting of the SS *Ponce* entering New York harbor was made to commemorate Marconi's radio coverage of the 1899 America's Cup races. At the request of the *New York Herald,* Marconi installed radio equipment on the ship and used it to cover the races, a stunning exhibition of the new technology.

In 1901, Marconi bridged the Atlantic when the single letter "s" was sent via Morse code from Poldhu, Cornwall, to St. John's, Newfoundland, a distance of nearly two thousand miles. This disproved prevailing theories that the curvature of the Earth would limit transmissions via electric waves to a mere one hundred to two hundred miles. It was a huge step forward in radio communications, and it set the stage for important new developments in the field of broadcasting and radio.

In May of that year, another radio milestone occurred: the *Lake Champlain*, a British cargo ship then owned by the Beaver Line, became the first merchant vessel to be outfitted with wireless. Marconi himself sent a radio message wishing Captain Stuart of the *Lake Champlain*, which was off the coast of Ireland at the time, "every success with the wireless system."[2]

Also in 1901, the first attempt to combine two fledgling lifesaving technologies—radio and lightships—occurred when Marconi oversaw the installation of a radio set on the Nantucket lightship, forty-two miles southeast of Nantucket Island. The village of Siasconset's new Marconi Wireless Station received its first signal from the lightship on August 12. In reality, the wireless set had been installed at the behest of the *New York Herald*, not so much as an aid to mariners but to speed up the process of reporting the shipping news. The lightship and Siasconset, on

Nantucket's east end, were both perfectly situated to communicate with the more than 250 ships passing daily, thus expediting reports related to shipping and, seemingly as an aside, to aid in any maritime rescues that might become necessary in the area. The experiment was deemed a success, and several other lightships would soon be equipped with radios.

In the meantime, Marconi continued experimenting, with the goal of increasing the distances at which signals could be transmitted and received. In July 1902, he succeeded in receiving signals on board an Italian warship, the *Carlo Alberto*, from Cornwall when the vessel was at the Russian port of Kronstadt, a distance of 1,600 miles; two months later, messages were received on the same vessel at Spezia, Italy, from Cornwall, a distance of over 800 miles across Europe *and* the intervening Alps.[3] The problem of distance, in and of itself, had seemingly been solved.

About this time, Marconi also laid to rest one of the early mysteries of radio, regularly receiving messages from distances of seven hundred miles by day and two thousand miles at night. This verified that messages can often be sent greater distances at night, because it turns out that sunlight ionizes part of the atmosphere, causing it to become partly conductant, thus decreasing transmission ranges; obviously, this does not happen at night. Marconi had theorized that this would occur, and he turned out to be correct.[4]

The year 1902 also saw another tremendous advancement in radio communications: Although discontinuous waves would satisfactorily transmit the dots and dashes of Morse code, high-quality voice and music cannot be transmitted in this way. So in 1902, Canadian American Reginald Aubrey Fessenden switched to using a continuous wave, becoming the first person to transmit voice and music by this method. It would take time to gain popularity—and clarity, especially over long distances—but it was clear that someday radio transmissions would involve not just Morse code but actual voices; perhaps even music could eventually be transmitted with a certain amount of fidelity!

By the following year, the use of Marconi sets on board ships and land stations was gaining traction. The Marconi system was installed and in use on some forty-five vessels, including liners from Cunard, the American Line, and the Red Star Line, and there were also fifty-four

In 1902, Canadian American inventor Reginald A. Fessenden became the first to transmit via continuous wave, opening up the possibility of using radio to transmit voice and music.

land-based stations in use by the end of that year.[5] (There were also competing systems offered by other manufacturers, but these never had quite the commercial impact of the Marconi apparatus.)

It's worth noting that these early radio installations were not without problems, and if you were contemplating a sea voyage, it might have been worth spending a few extra dollars for a cabin located as far away from the ship's radio shack as possible. The onboard transmitter received its power from a large bank of storage batteries; according to an official history published by the Marconi Company, the "crash of the spark in the open gap was a source of continual annoyance to those cabin passengers so unfortunate as to be quartered near the wireless room."[6]

The first time that the new radio gear proved its worth may have been in December 1903, when the ocean liner SS *Kroonland*—also built by Cramp and Sons, the same company that had built *Valencia* some years before—bound from Antwerp to New York, suffered a breakdown in her steering gear thirty miles from the Fastnet Lighthouse off of the remote Fastnet Rock in the Atlantic Ocean, the most southerly point of Ireland, lying about four miles southwest of Cape Clear Island and about eight miles from County Cork on the Irish mainland. The vessel was fitted with Marconi apparatus, and communication was immediately established with the station at Crookhaven, through which the captain of the vessel sent messages to his owners in Antwerp, receiving instructions in return, and a large number of passengers sent reassuring messages to their friends in all parts of the world.[7]

The final radio-related item of note here deals not with Marconi, but with Marconi's archrival, American inventor Lee de Forest, who in 1906—the same year that *Valencia* hit the rocks off the coast at Vancouver Island, British Columbia—invented what would come to be called the vacuum tube. The device, which remained in production in various

The SS *Kroonland*, shown here in a painting by Antonio Jacobsen, was en route to New York from Antwerp when, during a 1903 storm, she became the first ship to issue a wireless distress call.

forms well into the 1970s, when it was superseded by the transistor—which had been invented in 1948 by a team led by William Shockley at Bell Labs—could, among other things, act as an amplifier, ensuring that radio waves could propagate much longer distances while using less power. The vacuum tube, which de Forest called the Audion, eventually made possible talking motion pictures, the electronic oscillator, television, and many other devices. Important though it was, this latest advance in the development of radio would come much too late to help the passengers and crew of the ill-fated *Valencia*.

Chapter 7

Maritime Misfortunes

Around this time, several events occurred that would play a part in *Valencia*'s future. In 1902, the Pacific Packing and Navigation Company sold the ship to the Pacific Coast Steamship Company, which used her mostly for the California–Alaska run. Interestingly—and tellingly—*Valencia* had not been a well-thought-of vessel, even upon her original arrival on the Pacific coast back in 1898. She was regarded as being too small and too open to the elements (the elements in the Pacific Northwest being much rougher and voyages significantly more taxing than what she had encountered during her Venezuela runs), which meant that travelers reckoned her a second-class vessel. Most also thought of her as underpowered, given her average speed of only eleven knots.[1] Her long bow, which had served her well enough on the Atlantic, merely made it difficult to see what lay forward of the vessel on the foggy Pacific coast. This would become a concern later, when *Valencia*'s captain was peering ahead into the darkness and fog, trying to determine what obstacles might be in the ship's path as he attempted to navigate near the perilous coast of what would turn out to be Vancouver Island.

In February 1903, the *Valencia* rammed into a small wooden steamer, *Georgia*, not far from the Seattle docks. In spite of the fact that *Valencia* was twice the size of the wooden vessel, the larger ship was severely damaged and had to return to the docks for repair. *Georgia*, meanwhile, continued her run, only returning for repairs afterward.[2]

A different ship, but one related to the *Valencia* tale, was involved in a much more serious incident the following year. In late February, with captain N. E. Cousins in command (more about him and his vessel shortly), the SS *Queen* caught fire and burned.[3] On her scheduled passenger run up the coast from San Francisco to Puget Sound, having departed San Francisco on Thursday, February 25, off Tillamook Head, Oregon, on the morning of February 27, 1904, fire broke out belowdecks and quickly enveloped the ship's stern. There were 218 people aboard and fourteen lives were lost. *Valencia* and her future captain were not involved in the incident. However, Oscar M. Johnson—who would soon become captain of the SS *Valencia*—had occasionally commanded *Queen* on previous trips, including one to Seattle, and in fact that was Johnson's *only* prior experience sailing this route, which turns out to be of great importance a couple of years later, when Johnson commands *Valencia* on her disastrous voyage to Seattle. Ironically, Cousins would at that time command *Queen* when she was called to the aid of the *Valencia*. How well Cousins handled that distress call is still being debated and will be discussed in some detail shortly.

In October 1905, in an incident that eerily presaged *Valencia*'s ultimate—and not too distant—fate, Captain Johnson ran his vessel aground just outside St. Michael, Alaska;[4] the crew had to move seventy-five tons of cargo onto another vessel before they could free the ship.[5] This has been variously said to have been Johnson's second or fourth wreck, depending on the source one consults.[6] There are no reports of any other ships having run aground at Broad Point, where Captain Johnson ran the ship onto the rocks, so it wasn't known as a particularly troublesome area. However, in Johnson's defense, it must be pointed out that a snowstorm was blowing at the time, and visibility was quite poor.

Valencia will soon meet her end off of Vancouver Island, but it's important to keep in mind that she is far from the only ship to sink in that area. Only one year before the *Valencia* incident, in December 1905, a three-masted sailing vessel, the *King David*, blew ashore near Bajo Reef, just off the west coast of the island. While the crew all made it ashore, seven crewmen died after being sent out in a lifeboat to find help. (Or, in any case, they were never seen again. Some sources maintain they could have been picked up by another ship.)[7] Bajo Reef is located near Nootka Island, which, while remote, is still accessible via trails to such settled

SS *Queen* at Port Townsend, Washington, following a 1904 fire at sea that claimed fourteen lives

areas as Kendrick Camp, Plumper Harbor, and Yuquot. Thus, the *King David* sailors were able to make their way to civilization and most would ultimately be rescued by the steamer *Queen*, a vessel we've encountered before, and one that will also play a part in the ultimately unsuccessful *Valencia* rescue attempts. (The *King David* sailors were lucky; *Valencia* will wreck on the rocks at a much more unforgiving location midway down the western coast of Vancouver Island.)

And in late December 1905, almost exactly a year before *Valencia* would strike the rocks, the steel-hulled sailing bark *Pass of Melfort* was driven ashore on Vancouver Island. There were no survivors.

In fact, there were at least *seven* sinkings or strandings (the latter is the term used by the Canadian and US governments to describe a ship accidentally running aground on offshore rocks) in the area in the same year that *Valencia* wrecked. There were also four in 1905 and another five

The Scottish steel sailing vessel *King David*, docked in an unidentified port

in 1907. Between 1900 and 1906, there were a total of at least forty-three such incidents in British Columbia, most of them near Vancouver, Victoria, or Barkley Sound.[8] It was and still is a dangerous place to sail.

As it happens, *Valencia* did not usually make the San Francisco–Seattle run. That was the route normally taken by her sister ship, the *City of Puebla*. However, the latter vessel was undergoing repairs in San Francisco in January 1906, so *Valencia* was drafted to take her place; it was an ill-fated decision that would result in a tremendous loss of life and untold suffering for passengers and crew alike. It would also thrust into the public eye a cast of characters who, due to their actions after the grounding, would either be lauded as heroes or castigated as cowards.

PART II

The Storm Strikes

Pride is at the bottom of all great mistakes.

John Ruskin, *Modern Painters, Vol. IV*

Chapter 8

"A Series of Ill-Advised Mistakes"

When the SS *Valencia* departed San Francisco, there were at least two prideful, influential men aboard, and their histories were destined to become intertwined. In the end, one of them would live and one would die.

In fact, some months later, the city itself would almost die, and news of the *Valencia* disaster would be eclipsed, not just in the Pacific Northwest but around the world, by the San Francisco earthquake and by the great fire that ravaged the city afterward. On April 18, while most of the city slumbered, the San Andreas fault ruptured, some 296 miles away. At 5:12 a.m., a foreshock was widely felt throughout the San Francisco Bay area and, some thirty seconds later, the full fury of the earthquake itself struck the city. The magnitude 7.8 quake destroyed buildings in San Francisco and, significantly, ruptured gas lines, downed power lines, and shattered water lines, all of which made it almost impossible to fight the many fires that erupted. The city burned for three days.[1]

But this was a few months in the future. For now, the bustling "city by the bay," with a burgeoning population of some four hundred thousand people, sparkled in the winter sun and people went about their business, unaware of the twin tragedies looming in their future. The scene at the docks would have been one of barely contained chaos, or so it would have appeared to the uninitiated, as last-minute passengers hurried to the ship, while supply wagons and carts jostled and rattled their way onto the docks, and stevedores shouted and gesticulated as they finished loading supplies for the relatively short journey to Victoria, Seattle, and beyond.

Most of the damage in San Francisco was actually caused not by the earthquake itself but by subsequent fires.

THIS PHOTOGRAPH OF THE CITY'S MISSION DISTRICT WAS TAKEN BY H. D. CHADWICK OF THE US WAR DEPARTMENT.

The *Valencia*'s captain Johnson would have been unperturbed, of course; to his eye, the frenzied bedlam would have seemed normal, the unremarkable racket and workaday goings-on of a well-oiled nautical and commercial machine, with all of the frantic activity aimed at getting the passengers and supplies loaded and the ship safely under way as quickly as possible. He'd seen this purposeful disorder many times before.

Captain Oscar Marcus Johnson was of average height and build, and he wore his hair in a careful center part, as was the fashion at the time, swept forward in a valiant but mostly futile effort to disguise his rapidly receding hairline. His long, aquiline nose jutted out from atop a bristling moustache that hung below his upper lip and drooped downward, almost obscuring his lower lip; spreading outward to either side, it threatened to curl upward in a manner that, viewed from a modern perspective, might have suggested a cartoon villain. It was not lavish enough to be waxed and curled, but it was luxurious enough to dominate his narrow face.

When Captain Johnson gave the order to cast off from the docks in San Francisco, around 11:20 a.m. on Saturday, January 20, the 108 or so passengers and roughly sixty-five crewmembers (the count is inexact and often debated)[2] anticipated a milk run—an easy three-day journey to Seattle, much of it within sight of the coast.[3] Johnson in particular may have been too casual about the trip and too dismissive of its potential dangers; after all, ships had been making this run for many years, and the voyage was almost always uneventful. This voyage would be different.

Johnson was an experienced sailor, but perhaps not as practiced a captain as might have been desired. Born in Norway in 1866, he'd been working for the Pacific Coast Steamship company for between twelve and fifteen years, depending on the source one consults.[4] The forty-year-old officer had joined the Swedish Navy at fourteen, serving as an ordinary seaman. Over the years, he grew to be a serious and professional mariner

This National Archives and Records Administration photo, taken by a National Park Service photographer, shows the wharves at San Francisco in the year 1900.

who neither smoked nor drank and who was devoted to his craft. It's clear that Johnson was a knowledgeable sailor and was gaining experience as an officer when he sailed *Valencia* away from the docks at San Francisco in the winter of 1906, but it's also obvious—and will become more apparent shortly—that, though he was a skilled sailor, he was *not* an experienced commander. He was about to embark on a journey during which he would make a series of ill-advised mistakes of the sort that no experienced captain would make. Those mistakes would cause the deaths of at least 136 people, including his own, and would forever stain his reputation.

Johnson was not the only headstrong man aboard *Valencia* when she left San Francisco. Along with his wife and two children, educator Frank F. Bunker had booked passage on the ship, expecting an easy sail up to and then through the Strait of Juan de Fuca to the bustling city of Seattle, where Bunker was due to take up a position as the city's assistant superintendent of schools. (The city and its school district were rapidly expanding. With Seattle's population swelling from about eighty-one thousand in 1900 to almost two hundred thousand by the time *Valencia* set sail, the school district desperately needed more schools and more educators to teach in and administer those schools.) Bunker's aspirations would soon exceed this appointment, but for now, this position was the culmination of his professional goals. He and his wife, Isabelle, and their two children, Dorothy and Frank Jr., had been looking forward to an invigorating—but safe—voyage to Seattle. Instead Bunker's wife and children would die in front of his eyes and Bunker himself would barely survive.

The crew themselves were a varied and polyglot bunch. Many of them had transferred from *Valencia*'s sister ship, the *City of Puebla*. This would present—or exacerbate—problems later, when many crewmen seemed not to understand their duties.

It's ironic, though understandable, that the exact number of crewmembers—and passengers, for that matter—cannot be determined. As noted, the official tally of crewmembers was sixty-five, but other sources have it as fifty-six. This may be a simple transposition error, but getting an exact count of souls on board is problematic due to the garbled initial reports, the rush to get the investigation concluded (the federal

Few photos of Frank Bunker survive. This one is taken from the *San Francisco Call*, when Bunker became superintendent of schools in 1908.

investigating committee commission established by President Theodore Roosevelt deliberated and took testimony for only fourteen days), and by the fact that many bodies were never recovered and some of those that were recovered could not be positively identified.[5] It's possible that there was some sort of up-to-date manifest on board, but if so, it went down with the ship.

Many of the crew remain unheralded, but some are known to us either via their later testimony or due to their heroics during the wreck, or both.

Peter E. Peterson was the second officer of the *Valencia* and the only surviving deck officer.[6] Peterson was eventually rescued by the *City of Topeka*, and it was he who testified to the investigators about the soundings that Captain Johnson took—and failed to take. Peterson's testimony carried a great deal of weight, partly because, in contrast to Captain Johnson, the second officer had "made this trip something over a hundred times," though never on *Valencia*. He had, however, many times crewed as second mate on the *City of Puebla*.[7] Keep in mind that this ill-fated voyage to Seattle was only Johnson's second such journey, so the second officer was vastly more experienced than the captain, at least in terms of this run. Peterson was actually not supposed to be on the *Valencia* at all; he had joined the crew as a last-minute replacement for an officer who had transferred to another vessel.

John Segalos (some have it as Joe Cigalos) was a Greek seaman who shipped as a fireman, responsible for tending the fire to keep the steamship's boiler and engines running. It was a difficult and physically demanding job. Segalos too was rescued by the *City of Topeka*, as he and eighteen others, including Peterson, drifted half-frozen in an emergency raft that was barely afloat. Segalos, in broken English, testified to investigators that at one point he dove into the freezing water in an attempt to swim a line to shore, but had to turn back about halfway.

Boatswain (pronounced "bo'sun") Tim McCarthy was another crewmember who survived.[8] The senior nonofficer of the deck crew, McCarthy left *Valencia* aboard lifeboat No. 5, the same boat that would be discovered floating in the bay some twenty-six years later. He had been belowdecks when the *Valencia* struck the rocks, and he testified about the chaos that resulted when the lifeboats were partially lowered and secured (for a time) at the railing at the saloon level. Captain Johnson had ordered the boats lashed at the rail and had *not* given the order to board or lower the boats, but his wishes were ignored in the tumult. The four forward boats were lowered to the deck, and passengers lunged toward them, putting a strain on the tackles and davits that held them. From McCarthy comes

John Segalos was a Greek member of *Valencia*'s crew. During the sinking, he dove into the water in a failed attempt to bring a line to shore.

some of the most disturbing testimony about the destruction of the lifeboats as they were dropped, "cock-a-bill" (that is, at an extreme angle), from the ship's deck and then quickly destroyed, bashing their occupants into the side of the ship or dumping them into the sea.[9]

Fireman Frank Ritchley (some sources say that he was actually a fireman's messboy) also survived.[10] His story is notable mainly due to his later testimony and because he became part of the Bunker party and traveled overland with the group after reaching the shore. In spite of the fact that Richley held an official position of at least some consequence on the ship, it's interesting to note that it was Frank Bunker, not Ritchley, who became the de facto leader of their group once they had made it ashore. This could have reflected Bunker's vastly superior education, as well as his somewhat oversized ego and his determination to be in charge of any group of which he was a part. Mind you, Ritchley probably owed his life to Bunker; there was a point at which Richley had been ejected from the lifeboat by a powerful wave, and Bunker pulled him back in. To his credit, this did not stop Richley from later arguing with Bunker about the best way to get help for the stranded vessel.[11]

We will hear more about all of these men in later chapters. For now, it's enough to know that these six—and the roughly 167 other crewmembers and passengers—were some of the main characters in the tragedy that was about to play out on the decks of the SS *Valencia* and on the nearby shore. Most of the crew and passengers would die in the wreck, but five of these six (Oscar M. Johnson would go down with his ship) would join thirty-two other survivors to offer riveting, and often disturbing, testimony as to what happened aboard the ship, who was at fault, and—heartbreakingly—why more was not done to save those aboard.

Chapter 9

A Cascade of Errors

The SS *Valencia* left the docks and entered San Francisco Bay about 11:30 in the morning of January 20, bound for Seattle and Juneau, Alaska. She was not overladen, carrying only about 170 or so souls on board a vessel designed and licensed to carry almost 290. She also carried cargo: some 1,400 tons of canned goods, vegetables—including the aforementioned cabbages—and wine.[1]

Her captain, Oscar M. Johnson, almost certainly would have glanced up and out toward the docks, hoping to catch sight of his wife, Mary, and perhaps get a glimpse of the couple's three-year-old child as the ship prepared to cast off. The two had a happy marriage, but Mary dreaded his long absences at sea, and Johnson missed his family when he was away. When he returned from his voyages, Mary would wave to him from their front window, from which she could see the wharves. But he would not return from this last voyage, and Mary would wait and fret, assuring reporters that Oscar was a strong swimmer and was almost certainly safe. She was wrong.

While Los Angeles, four hundred miles to the south, was being inundated by torrential rains, the weather in San Francisco was fine as *Valencia* set sail, warm for January and mostly sunny. According to the *San Francisco Call* ("The Call prints more news than any other paper published in San Francisco," trumpeted the paper's masthead), "fair weather beckons" after a light frost the prior evening, with light north winds predicted, changing to southwesterly later in the day.[2] The ship set out on relatively calm seas for the Strait of Juan de Fuca, 660 nautical miles away. If the weather did turn, Johnson's options were limited, as he was well aware; there were few safe harbors along the way that were large enough to shelter a ship the size of *Valencia*. And finding safety by running the ship up on a sandy beach was unlikely because there *were* no sandy beaches: the shores along the route were rugged and consisted mostly of sharp, craggy rocks that could rip a ship's hull—even an iron hull—to pieces in minutes.[3] Johnson would soon be reminded of this, much to his sorrow.

For the first several hours of the voyage, all went well. *Valencia* steamed steadily into the night until, around 5:00 a.m. on Sunday, the lookouts, and undoubtedly some passengers, glimpsed the lighthouse at Cape Mendocino. So far, the ship had traveled some 190 nautical miles without incident and had been making good time.

As it turns out, the sight of the Cape Mendocino light was the last clear view of the shore afforded to the passengers and crew of *Valencia*. Soon afterward, fog closed in and the weather deteriorated. In theory, the sight of the Cape Mendocino light should have indicated that the *next* thing the officers and crew would see would be the Cape Flattery light, indicating that a turn eastward was in order so that the ship could enter the Strait of Juan de Fuca.

But Johnson never saw the Cape Flattery light. He did *claim* to have seen the Cape Blanco light, roughly midway between Port Orford and Bandon, Oregon, sometime later, and noted the supposed sighting in the logbook. Its presence would have meant that *Valencia* had now traveled 335 miles from San Francisco. However, no one else on board could confirm that sighting, and second officer Peter Peterson would later testify that he had definitely *not* seen the Cape Blanco light. (In the dry, measured words of the investigating committee's report, "Mr. Peterson was of the opinion that the entry was largely conjectural."[4]) This is the first indication we have that some of the ship's officers may have been at odds with the captain over his navigation, and disagreeing with his officers—many of whom were more experienced than he—may have been the first of Captain Johnson's several mistakes.

On Monday morning, the fog, wind, and stormy seas persisted, and visibility had been reduced to a couple of miles.[5] The crew was getting

The current Cape Mendocino light, in Shelter Cove, Humboldt County, California. It was moved to its current location in 1998. This was the last lighthouse the *Valencia* passengers and crew would see.

anxious, as was their captain. At this point, they should have been on a course that would have soon revealed the *Umatilla* lightship, but Johnson was beginning to wonder if they would raise that vessel. If sighted, the *Umatilla* lightship was a sign that Cape Flattery was fourteen miles away, indicating that the eastward turn into the Strait of Juan de Fuca was imminent. Peterson checked the log (not the logbook, but the *log*, a device that trailed behind the ship to measure the distance covered by the vessel) and decided that they were in fact near the lightship and would pass it late that evening.

Captain Johnson and his first mate, William Holmes, disagreed with Peterson.[6] They felt that the log was "overrunning," that is, that the ship was actually traveling more slowly than the log indicated. Johnson felt that the log was overrunning about 6 percent.

This was Johnson's second—and possibly most serious—error. There was only one possible reason for such a blunder: Johnson was unaware of the Davidson Current. Or, more precisely, he was not aware that the current, which even in 1906 had been well studied, was what oceanographers call *periodic*: it changed strength and direction seasonally, flowing mostly to the south in the summer but running, much more vigorously, to the north during the winter. Johnson had only made this run once before, and that was during the summer, when the current was minimal and flowing mostly in a southerly direction. He should have known that the current changed direction and strength during the winter months and was now propelling him directly toward the rocks off of Vancouver Island, British Columbia. (There were charts of the area that showed the Davidson Current and pointed out the seasonal changes, but Johnson—for whatever reason—did not have those charts aboard *Valencia*.)[7]

At this point, Johnson further demonstrated his incompetence (there is simply no other way to characterize his actions) by making two more mistakes. Ordering a series of soundings, he had a crewman drop a line with a lead weight at the end of it to determine the water's depth and discovered that the depth of the water beneath the keel was rapidly becoming shallower. Within an hour, the depth went from 80 fathoms (480 feet) to 60 fathoms (360 feet). The bottom was rapidly approaching. Because there is a steep drop-off at the entrance to the Strait of Juan de

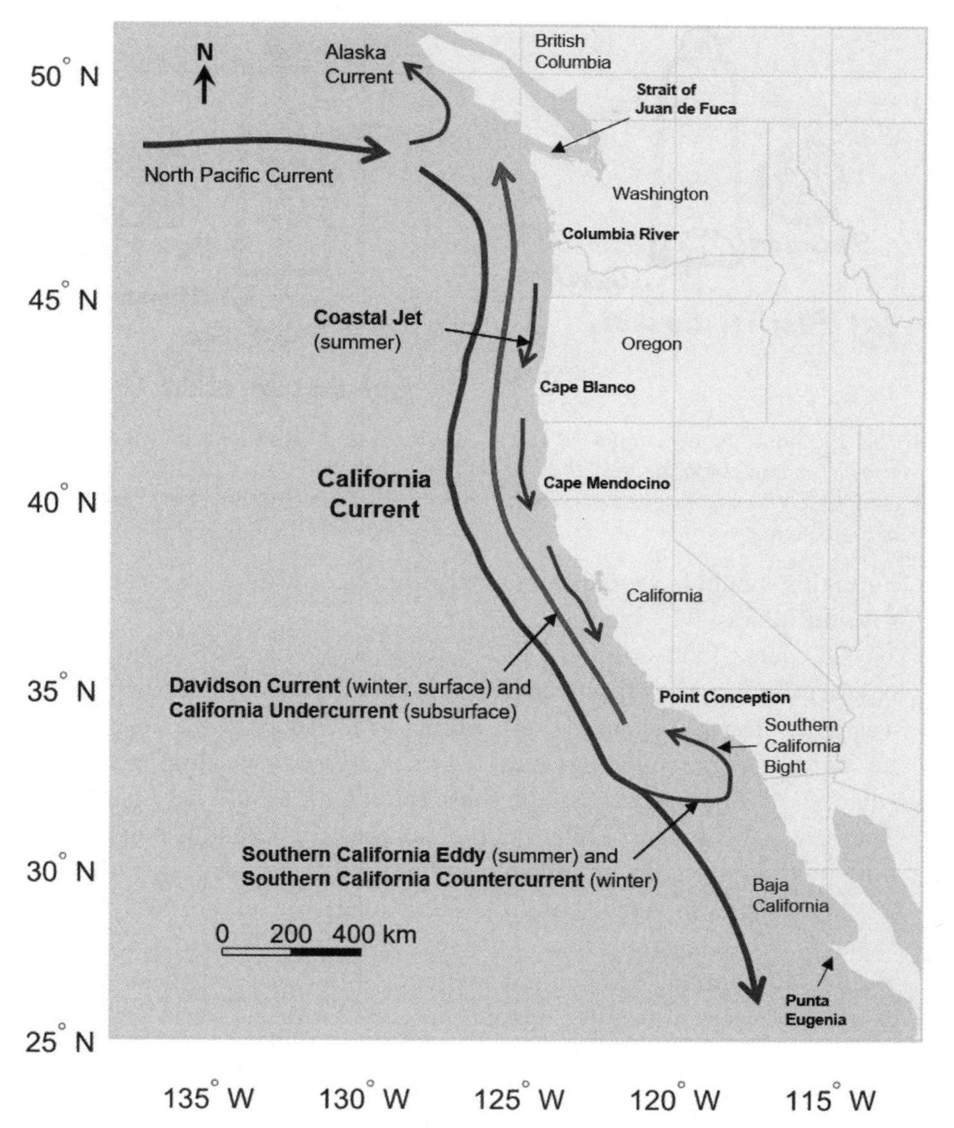

The Davidson Current is part of what is known as the California Current. As noted on this map, the Davidson Current tends toward the surface during the winter months, flowing directly toward Vancouver Island.

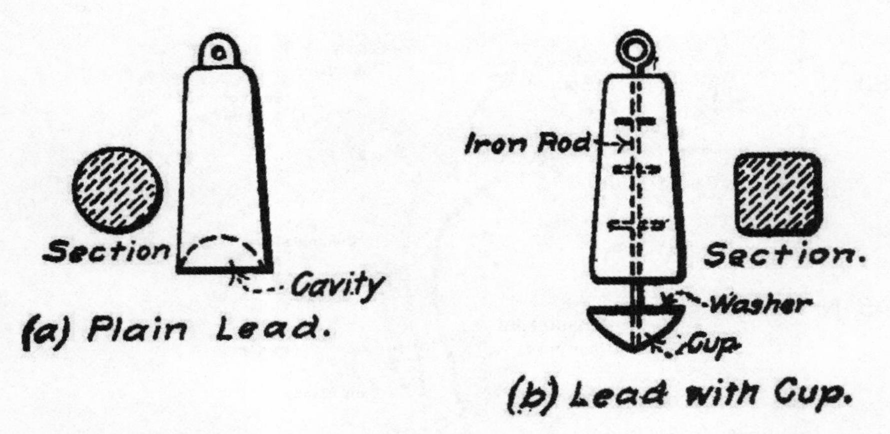

This 1905 drawing by Samuel Hill Lea shows two different types of sounding leads, one with a cup below the lead and one with an indentation or cup built in to the lead itself. Wax was inserted into the cup or indentation in order to retrieve samples of the bottom.

Fuca, this should have told Johnson that he was nowhere near where he thought he was.

Of course, Johnson should have ordered soundings earlier and more regularly. In fact, the official report on the investigation into the incident was quite clear about its findings relating to his soundings, noting that he "did not begin soundings until 6 o'clock Monday evening" and that what he did discover after taking those soundings "should have put him intensely on his guard."[8] Deferring the soundings until that Monday evening was a grievous sin of omission and indicative of poor seamanship.

Johnson's next blunder was also something that he did *not* do: the captain's refusal to simply turn out to sea and wait for daylight and better weather to continue his journey confounds modern researchers, much as it did the committee investigating the accident in the weeks after the grounding. Seafarers, like pilots, know that "sea room" (or, for a pilot, altitude) equates to safety. Danger lies near shore, where one can be driven upon the rocks; a ship is safer at sea, as an airplane is safer at altitude, where both have time and room to maneuver and where, if necessary, a ship can simply heave to or lie ahull until danger, in this case a stormy darkness, has passed.

An intelligent and prudent skipper would have turned and headed back out to sea. In fact, only hours after *Valencia* steamed onto the rocks, the captain of the SS *Edith*, following roughly in *Valencia*'s tracks and having run into essentially the same conditions, did the intelligent thing: he turned the vessel around, headed out to sea, and made it safely to Victoria the next day.[9] Discretion would have been, at this point, the better part of valor, but Captain Johnson failed to make the prudent move. Was he anxious to adhere to a strict timetable? Did he not wish to admit to his officers that he had erred? Was he simply paralyzed with indecision? We'll never know.[10]

Around 11 p.m., Johnson ordered the engines ahead dead slow, possibly thinking that he was nearing Cape Flattery, though he saw no light. It's also possible that he was, belatedly, realizing that he had no real idea *where* he was and felt that he should proceed with caution in the darkness, rain, and swirling fog. As it happens, he had earlier steamed right past the light, never hearing a foghorn blow, probably because, at the light itself, there was no fog. Even at dead slow, the current was propelling him quite quickly toward the rocky western shore of Vancouver Island, British Columbia. Moments later, Johnson was finally aware that his ship was in serious danger.

Chapter 10

Politics and Parsimony

Lighthouses have been around for a very, very long time. For many years, of course, people would simply build bonfires on rocks near the ocean as a way of warning off vessels sailing too close to the rocks, but the bonfires were lit only sporadically, and they were sometimes also confused with beacons meant to draw the ship in *closer*. One can see how this may have led to the very disasters the fires were meant to warn against.

The first actual lighthouse may have been built by the Egyptians around 280 BC and, at 450 feet high, it was also said to be the *tallest* lighthouse ever built. It was called the Pharos of Alexandria, and it was one of the Seven Wonders of the Ancient World, until it was damaged by a series of earthquakes and then abandoned after 1323 AD. In 1477, its remaining stones were used to build a Muslim fortress on the eastern side of the northern tip of Pharos Island.[1]

The first *modern* lighthouse, at least in the United States, was built in 1716 in Boston, Massachusetts. Boston was, after all, a busy harbor, and commerce demanded some safeguards. At the turn of the eighteenth century, the number of lighthouses skyrocketed, mainly due—again—to the increase in commerce, transatlantic and otherwise.

On the West Coast of the United States, though, even after coastal cities had been founded and settled, lighthouses were—and remained for many years—few and far between.

There were some, though. At Cape Flattery, near Neah Bay in Washington State, a lighthouse was built in 1854 on Tatoosh Island and illuminated in 1857. Now no longer manned, it was the northwesternmost

A view of the original Boston light, drawn by William Burgess in 1729. Note the signal cannon in the background.

lighthouse on the West Coast of the contiguous United States. This is the lighthouse that Captain Johnson and his crew *should have* spied as they approached the entrance to the Strait of Juan de Fuca; unfortunately, in 1906, it was one of very few lighthouses in the area.

It's not always possible to build a lighthouse exactly where needed, but it *is* often possible to station a *lightship* near navigational hazards to warn off ships in the vicinity. In 1898, *Light Vessel No. 67* became the first such ship on the Washington coast, stationed northwest of Cape Flattery. A second light ship, *Light Vessel No. 93*, did not appear on the Washington coast until 1909, three years *after* the sinking of the SS *Valencia*.[2]

Between the two, lightships and lighthouses, the United States made at least some effort, often augmented by state governments and local municipalities, to ensure the safety of mariners plying the waters off both coasts of the country.

Fisgard lighthouse on Fisgard Island, BC, erected in 1860, was the first lighthouse built on the west coast of Canada.

PHOTO BY LESLEY SCHER. USED WITH PERMISSION.

Lightship No. 51 at Sandy Hook, New Jersey, pictured in the 1890s

The same could not always be said for the Canadian government, especially when it came to the far-flung, remote regions of British Columbia. As author and former Canadian Coast Guard Superintendent Clay Evans points out, it was not until the loss of the *Valencia* that "the long-standing cries for more lifesaving measures at the entrance to the Strait of Juan de Fuca and up the west coast of Vancouver Island reached a crescendo amongst the press and the populace."[3] Although the area was a strategic shipping choke point, Evans points out, "there were no other safety measures" nearby.[4]

The unfortunate fact was that Vancouver Island, and especially its west coast, was considered a peaceful, rustic backwater, deserving of very little oversight and even less funding. Even after the loss of the *Valencia*, the Canadian Department of Marine and Fisheries (an organization that would eventually become the Canadian Coast Guard) would, while boasting in its Annual Report that the organization had ninety-eight lifesaving stations in Canada, fail to mention that not one of those was on the Pacific coast of the nation.

The absence of lifesaving infrastructure on the west coast of Vancouver Island had little to do with necessity and everything to do with politics and parsimony. The risks had been known for many decades; this was a volatile coast rent with violent storms and littered with innumerable wrecks. But at the time, any changes—including enhancements to maritime safety—had to go through the governor in distant Ottawa. This was especially true of enhancements that cost a great deal of money, as most of those dealing with the safety of mariners surely did.

Ironically, many heroic rescues were successfully undertaken by the supposedly "savage" Indigenous inhabitants, even in the absence of any official government infrastructure, and almost always absent any form of remuneration. The First Nations people were, after all, superb seamen and adept at launching and retrieving boats in the wild surf off the coast of Vancouver Island and elsewhere. At one point, Colonel W. P. Anderson, the man in charge of lighthouses and lifesaving stations in the Dominion of Canada, even proposed that the native people of the west coast be drafted into rescue teams and paid a (one assumes nominal) salary. This was right after he pointed out that it was "inexpedient to establish a life-boat [on the site then in question] at least until the danger becomes more urgent or the white population denser."[5]

In the end, little was done by the government of Canada to provide help for endangered mariners near the west coast of Vancouver Island, even after an 1857 gold rush exacerbated the problem by ensuring that thousands of foreign-flagged ships would be pouring into the port of Victoria, thereby almost certainly increasing the number of wrecks occurring on or near the coast of the island.

This was the bleak situation into which Captain Johnson sailed the SS *Valencia* in January 1906. There was one lighthouse at Cape Flattery, but he had not sighted it and was now well north and west of it. There were few lighthouses on the west coast of Vancouver Island itself and no lightships stationed nearby to warn him that he was rapidly approaching the razor-sharp rocks that lined the coast near Pachena Point, British Columbia. The closest other lighthouse was the one at Cape Beale, about

Col. W. P. Anderson, pictured here in 1904, was a Canadian civil engineer who for many years was in charge of lighthouses and other lifesaving infrastructure in Canada.

COURTESY OF LIBRARY AND ARCHIVES CANADA, REPRODUCTION REFERENCE NUMBER E010752608

ten miles distant in the fog and mostly concealed by the storm and by the promontory of the cape itself. Because it was hidden by both geography and weather, there was nothing to warn Johnson and his crew that they were speeding toward almost certain death.

Chapter 11

"A Bright Star amid the Primeval Gloom"

The Cape Flattery light that signaled the easterly turn into the Strait of Juan de Fuca was one of few aids to navigation in the area in 1906, and it was greatly relied upon by sailors of the era. In fact, before the lighthouse had been built, ships approaching the Strait of Juan de Fuca at night were simply advised not to attempt to enter the strait at all; in the dark, it was too easy to miss the turn or to turn too sharply, risking running aground on the rocks and shoals surrounding Cape Flattery. Instead vessels typically stayed well out of the mouth of the strait, remaining at sea until daylight, when they could see to navigate. After the construction of the Cape Flattery light, sighting it was surely a relief to tired sailors who may have spent many days at sea.

The lighthouse had a bit of a checkered past, though. It was begun in 1854 and completed in 1857, but the builders ran into constant difficulties with the Indigenous inhabitants of the area; before they could even begin construction of the lighthouse, the builders found it necessary to build a blockhouse and to furnish it with muskets and ammunition to be used for protection from the Indians.[1]

The Cape Flattery light, on Tatoosh Island northwest of Neah Bay at the entrance to the strait, featured a first-order Fresnel lens and was augmented by the 1872 addition of a twelve-inch steam whistle that was used as a foghorn.[2] Once the lighthouse, a conical tower made of sandstone and brick, was complete and activated, it served as a beacon to protect potentially wayward vessels. However, it also turned out to be a popular gathering point for neighboring natives, who found that the

The Cape Flattery lighthouse, on Tatoosh Island, overlooks Cape Flattery and guards the entrance to the Strait of Juan de Fuca.
NOAA

lighthouse made an excellent fishing and whaling station. This so irritated and frustrated the first lighthouse keeper that he resigned soon after the light was finished.[3]

On the other hand, the natives had good reason to be hostile. While the US government had paid some $30,000 for seventeen acres of the Makah tribe's traditional lands, an 1853 smallpox outbreak had angered the tribe, making the tribal elders reluctant to countenance the construction workers' presence. The strained relationships with the locals—combined with poor pay and dismal conditions at the site of the lighthouse—resulted in the resignation of not just the first keeper but of several successive keepers.[4]

The name of the cape itself comes from Captain James Cook, who was exploring the waters off the tip of the Olympic Peninsula and who claimed that an opening along the coast "flattered" him into believing that he had found an important harbor. Oddly, Cook doubted the existence of the Strait of Juan de Fuca, noting in his logbook: "In this very latitude geographers have placed the pretended Strait of Juan de Fuca.

But nothing of that kind presented itself to our view, nor is it probable that any such thing ever existed."[5]

The lighthouse's steam-powered fog signal, built well after the lighthouse itself, was housed in a separate building. Ironically, activation of the whistle was delayed due to a lack of water: although built in the summer of 1872, the whistle could not be utilized until November of that year, after the rainy season had begun, when the cistern supplying water for the steam that would power the whistle had filled adequately. As it turned out, that signal was replaced in 1897 by a new brick signal building, after several years of such extreme disrepair that the original signal could not be counted upon for use when needed.

At the time of the *Valencia* incident, the keeper at the Cape Flattery light was a man named John Cowan, who had arrived in 1900 with his wife and seven children. The family sent the children off to Portland, Oregon, so that they could attend school, and it is said that the Cowans never once left the island during that period.

Cowan, who served as keeper at the Cape Flattery light for over thirty years, is credited with saving several lives while there, but his son Forrest—serving at the time as Cowan's assistant—was lost at sea during one of the rescues. Two other lighthouse keepers working at Cape Flattery, Nels Nelson and Frank Reif, lost their lives while attempting to leave the island during a storm. Their bodies, like some of the victims of the *Valencia* tragedy, washed up on a Vancouver Island beach over a week later.[6]

The life of a lighthouse keeper was a difficult and dangerous one, especially back then. For his dangerous and backbreaking work, Cowan's salary when he began working at Cape Flattery in 1900 was a princely $900 per year; when he retired in 1932, it had risen to $1,320 per year.[7]

Understandably, the lighthouse was a much-appreciated addition, welcomed by navigators of all stripes. After many years of sailing blindly up the coast, the presence of the bright light must have been a tremendous comfort to ships in the area. As author/explorer James Swan put it in 1859, and as quoted by George Nicholson in *Vancouver Island's West Coast 1762–1962*, "The wind continuing adverse, we were obliged to beat across the entrance of the strait (Juan de Fuca) for five days

Lighthouse keeper John M. Cowan in 1900, during his days as keeper of the Cape Flattery light

without gaining anything. But every night we were cheered by the light of Tatoosh Island, which shining like a bright star amid the primeval gloom, civilization and a proof that the 'star of empire' had made its way westward till the waters of the Pacific had opposed a barrier to the tide of emigration."[8]

The Cape Flattery lighthouse is now electrified and is fitted with a diaphone fog signal and a radio beacon.[9] It has been automated since 1977, so for its last years of operation, no full-time keeper was required. Standing 165 feet above the water, the light can be seen for nineteen miles. It was automated in 2008 and turned over to the Makah Indian Tribe, which controls the island.

Valencia never sighted the Cape Flattery light. Not only that, but those aboard *Valencia* also never heard the Cape Flattery light's foghorn, even though the ship had encountered a tremendous amount of fog. One possible explanation is that the vessel was too far away from the foghorn to hear it—after all, the ship was now several miles west of the lighthouse itself. More likely, though, is the possibility that the foghorn was simply not activated, largely because there *was* no fog near the lighthouse; while *Valencia* may have been engulfed in a thick fog, the weather was reported to be clear at Cape Flattery.[10]

Chapter 12

The Deadly Island

With an area of some 12,000 square miles and a length of over 280 miles, the island toward which *Valencia* was unknowingly steaming is not simply large, it is *huge*—the largest island on the Pacific coast of North America. It is, in fact, the top of a mostly submerged mountain range, and the island formed by the peaks is separated from mainland Canada by the Georgia, Queen Charlotte, and Johnstone straits.[1] The island is separated from the United States by the Strait of Juan de Fuca. *Valencia* should have made an easterly turn on sighting the Cape Flattery light and headed up the Strait of Juan de Fuca, traveling between US territory on her starboard side and Canadian on her port; instead she missed that turn and continued, unknowingly heading toward the rocky west coast of the island.

The island is heavily wooded and mountainous, its coast lined with fjords, dotted with small bays, inlets, and sounds, and furrowed with rivers—some of them quite turbulent—that empty into the sea, sometimes after extending for miles into the interior. It's very rough topography, even today a mostly unspoiled wilderness that would be difficult or impossible to traverse on foot. One cannot easily walk—or even drive—across the island from one coast to the other, even though the island is only sixty or so miles across at its widest point, and in 1906, large portions of the island were simply impassable.[2]

The island's west coast is rugged and, compared to the east coast of the island, sparsely populated. Especially back in 1906, when *Valencia* crashed into the rocks off its shore, there were few people and no roads or other

Vancouver Island is much larger than most people think, stretching some 280 miles from end to end and spanning roughly 60 miles across at its widest point.
IMAGE PLACED IN THE PUBLIC DOMAIN BY NIKATER, BACKGROUND MAP BY DEMIS, WWW.DEMIS.NL

infrastructure. Today, a modern highway, Highway 19, runs along the east coast, stretching all the way from Port Hardy on the north to Nanaimo and then via the Trans-Canada Highway 1 to Victoria on the south. Along the way, one frequently encounters quaint, bustling villages and busy towns, such as Campbell River and Nanaimo, that hum with commerce generated by tourists, fishing boats, and other industry. But even now, no highway runs along the entirety of the west coast of the island. One, Highway 14, runs for a short distance, between Sooke and Port Renfrew. Instead a patchwork of trails—with the Pacific Rim National Park's West Coast Trail among the most well known—is just about the only way to get from the northern to the southern tip of the west coast. In 1906, of course, the demanding coast was even more difficult to travel.

Interestingly, the island is constantly *moving*. Every year or so, the island moves about one centimeter closer to the mainland, due to small tremors. In fact, the island moves back and forth a bit, but overall, most of the movement is to the east.[3] This means that Vancouver Island *could* someday collide with the mainland, but that would be many millions of years in the future, if it happens at all, and we most likely will have other issues to contend with by then—if we're around to worry about it at all.

The climate of the island is mild, compared to that of mainland Canada. It rains often during the autumn and winter months, and the temperatures along both coasts are consistent, with mild winters and relatively cool summers. It rarely snows at low altitudes, but snow is common on the mountaintops, leading to a thriving, though small-scale, ski resort industry.[4] This brings up a question: Why, if the climate tends to be mild, is the ocean so turbulent—and why are shipwrecks so common—on the western coast of the island?

The answer has largely to do with the topography of the coast. Recall that the island is in fact a mountaintop. Lying just off the coast are other rock outcroppings; on land they would simply be smaller mountains associated with the larger ones, but submerged they become dangerous reefs and crags that can rip the bottom out of a ship within minutes. The west coast of the island is also exposed to moisture-laden winds; when the winds approach the island, the warm air masses are forced upward, resulting in heavy precipitation. Because the coastline is exposed, it can be struck by storm systems that originate far offshore and that often gain strength as they approach the island. Especially in winter, approaching frontal systems can provide for dramatic—and dangerous—weather. Meanwhile, the east coast of Vancouver Island is somewhat shielded by the landmass of mainland Canada. The result is an island the west coast of which can be lethal to unwary or unprepared seagoers, as *Valencia*'s captain, crew, and passengers are about to find out.

Most history books begin the story of Vancouver Island by recounting the explorations of Captain James Cook and George Vancouver, noting that the island was held by the Hudson's Bay Company until it was made a British crown colony in 1849.[5] It became part of British Columbia in 1866, and British Columbia itself entered the Dominion of Canada as a province in 1871; Victoria, on the southern tip, was declared its provincial capital.[6]

This approach to the history of the island is accurate, as far as it goes, but it leaves out—or at the very least is dismissive of—some thirteen thousand years of exploration and habitation by various Indigenous tribes that made the island their home long before the Europeans arrived. Those tribes have inhabited the island since the Ice Age, and their impact on the island's culture is still vivid, though it was minimized for many, many years, even extending to various attempts at forced assimilation. Unfortunately, as in other areas of Canada and the United States, it was for many years official government policy to remove young Native children from their homes and send them to residential schools, the express purpose of which was "killing the Indian in the child." That is, the intent was to churn out assimilated, some might say *homogenized*, citizens of the country, whether US or Canadian, while having—often forcibly—removed their Native identities. In Canada, some 150,000 children were taken from their homes and sent to these so-called residential schools.[7]

There are multiple First Nations groups on the island: the largest include the Coastal Salish on the south and east coasts the Kwakwaka'wakw mostly in the central and northeast portions of the island, and the Nuu-chah-nulth on the west coast. As with Native American peoples in the United States, each tribe has its own art, spiritual culture, and architecture; largely because of the readily available food supply afforded by the ocean, rivers, and lush woodlands, the various Native peoples on the island had the time and leisure to develop a rich culture that included art, crafts, storytelling, and religious and social traditions. Surrounded by the sea, they developed an intimate relationship with the waters around them and with the forests that grew thickly almost down to the water.[8] This all occurred thousands of years before the island was "discovered" by Europeans, so to begin a retelling of the island's history with stories of George Vancouver and Captain Cook is misleading at best and minimizes the contributions of the First Nations people who lived there.[9]

Today, many towns and cities on the island exhibit a distinctly First Nations character, often commingled with the legacies of the area's Spanish and British influences, thus reflecting the island's varied history.[10] With sentiment growing to find ways to build new relationships between Indigenous peoples and other Canadians, many people on the

This statue of George Vancouver stands in front of Vancouver City Hall, in Vancouver, British Columbia, Canada.

IMAGE PLACED IN THE PUBLIC DOMAIN BY USER ZHATT

A contemporary deer hide beaver drum made by artist Richie Brown, a member of the Snuneymuxw First Nation

PHOTO BY LESLEY SCHER, TAKEN AT THE NANAIMO MUSEUM, NANAIMO, BRITISH COLUMBIA, CANADA. USED WITH PERMISSION.

island and elsewhere are seeking to ensure that the rights of First Nations people are recognized and that reconciliation is sought for earlier grievances, including loss of land, language, and culture. Many of the towns cater to tourists from the mainland and around the world, but these days their offerings often include not just Native arts and crafts but also T-shirts and other items extolling the virtues of "truth and reconciliation" and imprinted with the slogan, "Every Child Matters."[11] The sentiment is belated, but no doubt heartfelt.

Chapter 13

Doomed, Alone, and Out of Contact

As the SS *Valencia* steamed toward her fate in late January 1906, neither she nor most other vessels her size carried any radio equipment, although such equipment, rudimentary as it was, existed. At the time, there *were* some ships that carried radios, but marine radio—indeed, radio in general—was still in its infancy. While a few dozen ships were equipped with Marconi sets by 1906, most of those were large transoceanic liners; *Valencia*, a small vessel employed for limited coastal runs, wasn't one of them.

Most of the larger civilian ships, such as the SS *Hamburg* and the SS *Deutschland*, both of the America-Hamburg Line, functioned—as cruise ships do today—as miniature cities, complete with game rooms, beauty salons, banks, gymnasiums, ballrooms, onboard entertainment revues, and the like. Some were large enough, in fact, to publish their own daily newspapers while at sea. The *Hamburg*'s was called *The Atlantic Daily News*, and—being a first-class, oceangoing liner—it included a section called "Wireless News," which recounted world news as received en route by the ship's radio system; while at sea, passengers could read fairly up-to-date news stories about Russian "terrorists" robbing a St. Petersburg van carrying customs receipts, or the fact that the Duke and Duchess of Marlborough had separated. (One issue printed a special notice requesting that first- and second-class passengers refrain from throwing money or food to the passengers in steerage, "thereby creating disturbance and annoyance.")[1] Well-heeled passengers could even send messages to friends and family back on land, although that was a very expensive novelty reserved for times when official radio traffic was at a minimum.

The news was delivered via what the line called "Special Marconi-grams," so, even at sea, marketing genius Guglielmo Marconi made sure that his name remained in the public ear.

When *Valencia* sailed for Seattle, Marconi remained in the forefront of radio experimenters, determined to make radio a commercially feasible reality. He had successfully transmitted a signal across the Atlantic and was working to get "Marconi sets" installed on land and sea stations around the world. In only a few months, Lee de Forest (see chapter 6), an American inventor and electronics pioneer, would invent what would come to be called a vacuum tube, a device that could act as both a switch and an amplifier; this would spur the development of radio (and lead

The SS *Hamburg*, one of the earlier ships equipped with radio equipment, led a long and varied life. Captured in New York at the start of World War I, she was chartered by the Red Cross to ferry medical supplies to Europe, and then became a US Navy vessel, renamed first the USS *Hamburg* and then the USS *Powhatan*. In 1920, she again became a private commercial vessel and was finally broken up in 1928.

directly to the invention of the transistor, and thus the integrated circuit and other such advances), but at the time of *Valencia*'s sailing, the vacuum tube was still months in the future.

It was a Canadian American, Reginald Fessenden, who would make the next important series of radio-related advancements. Just prior to *Valencia* setting sail, Fessenden had made the first successful two-way radio transmission, from Brant Rock, Massachusetts, to Machrihanish, Scotland. He also developed a way to transmit audio signals over radio waves, which would eventually lead to commercial radio broadcasting. Fessenden also claimed to have broadcast the first commercial radio program even as *Valencia* steamed toward her fate. That broadcast was said to have included a speech by Fessenden, some recorded music played on a phonograph, and a reading from the Bible. However, he did not mention this accomplishment until 1932, and a lack of verifiable details has led to some skepticism about the claim. A few years later, during World War I, Fessenden would work on the development of sonar (sound navigation and ranging) as a tool to be used for submarine detection.

A few years after the *Valencia* was wrecked, the sinking of the RMS *Titanic*, in April 1912, would serve to underscore the importance of having radio installed in ocean liners. Some 31 percent (705 or so out of 2,207 souls) of the *Titanic*'s crew and passenger complement would be rescued, compared to only 21 percent (37 out of 173 or so) of *Valencia*'s, partly because the radio on board *Titanic* had been used to send a mayday message and request urgent help after the ship had hit an iceberg. This may be a bit misleading, though, given the dire situation in which the *Valencia*'s crew and passengers found themselves: so many on board *Valencia* succumbed so quickly to hypothermia or were drowned in the surf, killed when their lifeboats were crushed, or beaten to death on the rocks that having a radio may not have helped a great deal. The statistics also seem to imply that, at least according to one point of view, the *Valencia* is the grimmer disaster of the two, given that a smaller percentage of souls survived. In any case, the main rescue ship, the *Carpathia*, received *Titanic*'s SOS signal and steamed at top speed, about seventeen knots, toward the scene of the disaster, arriving about three hours after the ship had gone down. It picked up all 705 survivors, most of them in lifeboats

FESSENDEN Wireless Telegraph System

NATIONAL ELECTRIC SIGNALLING COMPANY.

Fessenden Wireless Telegraph Station Working between New York and Philadelphia.

As the result of five years' experimental work, including a working test of a full year, this system is now put on the market as being equal as regards speed and reliability to manually operated wire lines, while first cost and maintenance are only a small fraction of that of wire lines.

This system uses no coherer, the receiver consisting of a minute cylinder of liquid, whose resistance is lowered by the heating effect of the electric waves. As it is approximately 25 to 50 times as sensitive as the Solari coherer, it requires less energy for given distances, and is admirably adapted to sharp tuning, and overcoming interferences and difficulties from atmospheric disturbances.

This system does not infringe the patents of any other company, and the operation of the apparatus is guaranteed.

Telegraphic sets for working up to 150 miles overland or to 350 miles over sea are now standardised and can be supplied from stock or on short notice. Sets for working over longer distances supplied at short notice. Sets can be tested by purchaser before delivery between the Company's test stations, approximately 90 miles apart overland, or between the Company's marine stations.

No expert knowledge needed, as any telegraph operator can handle after a week's practice.

NATIONAL ELECTRIC SIGNALLING COMPANY,
WASHINGTON, D.C., U.S.A.

By the early 1900s, Canadian American inventor Reginald Fessenden was attempting to compete with Marconi and others through his National Electric Signalling Company. The effort failed, largely due to his insistence on charging higher prices than Marconi.

floating near the scene. (Other ships also responded, but they were too far away to do much good.)

For the *Titanic*'s owners (the White Star Line, actually a subsidiary of an American company, J. P. Morgan's International Mercantile Marine Co.), equipping their liners with radio was good business, but since 1910 in the United States, vessels carrying more than fifty passengers or traveling more than two hundred miles had been *required* to carry both a radio and a trained radio operator.[2] One year after the *Titanic* went down, the International Convention for the Safety of Life at Sea produced a treaty that not only mandated shipboard radios but also required them to be manned twenty-four hours a day.

As effective as radio was during the sinking of the *Titanic*, it turned out that radio interference had nonetheless hampered rescue efforts.

The White Star Line's RMS *Titanic*, shown here departing Southampton on her ill-fated maiden voyage, truly was titanic. At almost 883 feet in length, she was the largest ship afloat at the time. Some seven hundred or more of her crew and passengers survived her sinking, largely because she was equipped with radio, which was used to send out a distress signal.

Further steps were thus taken to avoid such issues, resulting in the US Congress enacting S. 6412, commonly known as the Radio Act of 1912, "An act to regulate radio communications." The act specified that all radio operators were to be federally licensed, much as today's ham and commercial radio operators are still required to be licensed, and that all ships must maintain a constant radio alert for distress signals, for which a separate frequency had been set aside. This marked the US government's first official entry into the regulatory realm of radio.[3]

The distress signal agreed upon by the 1906 International Radio Telegraphic Convention was "SOS," which does not stand for "save our souls" or "save our ship" but is simply an easily remembered and easy-to-transmit series of readily discernible sounds. As it happens, the *Titanic* was not the first ship to send a distress signal via radio. In 1909, both the Cunard liner RMS *Slavonia* (in June) and the steamer SS *Arapahoe* (in August) sent distress signals that resulted in ships sailing to their rescue; as far as we know, these were the first two ships to have transmitted "SOS" as a distress call, roughly three years before the sinking of the *Titanic*. (An alternative signal, "CQD," was also popular for a time but its use died out quickly after the sinking of the *Titanic*.)[4]

Of course, none of this did the *Valencia*'s passengers and crew any good. The ship steamed toward her dark destiny with no way of communicating with the world around her. From the moment she left San Francisco, she was doomed, and she was alone.

Chapter 14

"The Officers Were Now Concerned"

Frank Bunker, in 1906, a tall, serious gentleman who looked younger than his thirty-three years, with thinning, sandy hair, and prominent ears, was a night owl. Late on the night of Monday, January 22, he left the family's small stateroom aboard the SS *Valencia*, heading up the passageway to the ladder that led to the deck above. He left his wife below, also awake and fully dressed, tending to their two children. There, around 11:00 p.m. or so, Bunker found the captain and crew taking soundings.

The taking of soundings was a procedure about which Bunker had read a great deal but had never seen, so he was keenly interested in the process. In those days, taking a sounding meant dropping a lead-weighted line over the side. The line was marked to indicate the depth of water beneath the keel. Allowances had to be made for the angle of the line, which varied depending on the vessel's speed and the scope (that is, length) of the line, but after those allowances had been made, a competent seaman could determine fairly accurately the depth of the surrounding water.

However, nautical charts do more than simply indicate the water depth, because the depth is not the only thing that matters; charts also note the makeup of the *bottom* in the area, because that might also help determine the vessel's location. The seafloor might be rock, weeds, mud, or some other material, and the composition of the bottom was and still is noted on many charts; the lead weight on the bottom of the line included a concave area meant to collect samples of the bottom. Into that concave area was inserted wax or another sticky substance; when lowered to the seafloor, the wax would pick up samples of the bottom. If the sounding line indicated, say, a mud bottom and a depth of twenty fathoms (a fathom is equal to six feet, so 20 fathoms is 120 feet), the ship's officers could check against the chart and those two pieces of information would tell them if the ship was where they thought she was. If there was no mud on the weight, but the chart indicated that there should be, then even if the depth seemed to match the chart, the fact that the bottom did not was an important—perhaps critical—piece of information. In this case, it was the depth that did not match the chart's depiction of where Johnson *thought* he was.

On that night, consternation was evident among the officers, but few passengers were around to notice. At 11:00, the sounding indicated a depth of forty fathoms. At 11:30, the bottom was only thirty fathoms

This is a sounding lead, angled such that you can see the indentation on the bottom. Wax or another sticky substance would be placed in the indentation, and dropping the lead to the seafloor and then retrieving it would bring up samples of the bottom.

COURTESY OF ED MORRIS, M&B SHIPCANVAS CO. USED WITH PERMISSION.

beneath the keel; it was rapidly shallowing. If the ship were near the entrance to the Strait of Juan de Fuca, where the captain had thought she was, the bottom should be much deeper.

The officers were now concerned, and that is when Frank Bunker stepped out on deck and took shelter from the weather in the ship's social hall, a room reserved for lounging, card playing, and, with the furniture cleared away, occasionally for dancing. The officers conferred and, at 11:45, dropped the sounding line again. This time the depth registered only twenty-four fathoms. Things were getting worse.

Bunker, however, was no seaman. He could see the crew taking the soundings and discussing the results, but, in his words, he "attached no significance to it . . . at that time."[1] Bunker remained unconcerned.

A few minutes later, Bunker had plenty of cause for concern. The ship, slowed to just about the lowest speed it could manage while still maintaining steerageway, struck a reef. The vessel shuddered, then, "only a few moments" later, according to Bunker, struck again and then again. That first encounter with what turned out to be Walla Walla Reef consisted of three quick strikes.

At the time, Bunker—and some other passengers who had come on deck—felt that matters could be put right simply by backing off of the rocks. Moments later, though, the ship appeared to be sinking, and Bunker heard the cry, "Take to the boats!"

At this point, it's important to note two things. First, Captain Johnson himself had not yet said that the crew should lower, and the passengers should board, the lifeboats. Shortly, the captain would order the boats lowered to the rail, but that's as far as he went; he did not yet order the boats loosed, boarded, and launched. Second, Johnson had been remiss in his crew training prior to and throughout the voyage. He never held a full lifeboat drill; thus, neither the passengers nor many of the crew were familiar with their posts and duties in the event of an emergency. Just as Johnson had failed to take soundings until it was too late to matter, he had also failed to prepare his crew and passengers to safely abandon ship until it was too late to do so. The ship was in the wrong place due to his incompetence and the passengers and crew were in peril because few knew how to safely abandon ship, a sad state of affairs that was *also* due to his incompetence.

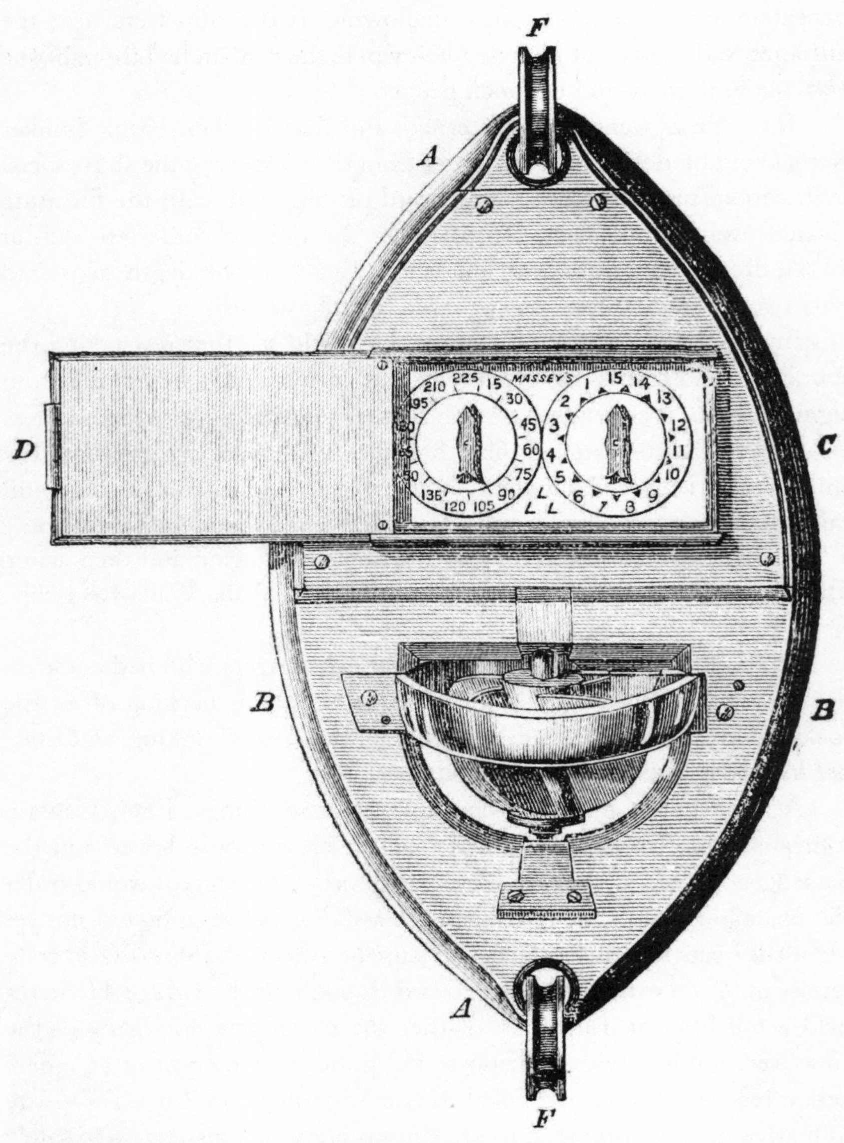

Some sounding devices were fairly complex. Edward Massey's "sounding machine," invented in 1802, featured a rotor connected to a dial. When the lead struck the seafloor, the rotor would lock, allowing the operator to read the depth from the dial.

Now that he knew that his ship was sinking, Captain Johnson finally acted intelligently. Freeing the vessel from the rocks, he pivoted *Valencia* around the reef such that her stern was pointed to the shore. His intent was commendable: He thought that perhaps he could force the ship to beach by running the engine full speed astern and putting the rear of the ship close enough to the coastline that the crew and passengers would be able to reach the shore. But there was no beach; instead the ship simply drove up onto the reef, leaving the vessel caught on the rocks.

Valencia was now doomed. With her steel hull rent by the huge rocks, water poured in, dousing the engines, and the vessel settled on the reef for good. The engines and generators having failed, the lights went out. Meanwhile, the relentless waves pounded on the ship, sometimes rising high enough to crash into and damage the spars and masts that rose far above the decks. *Valencia* was quickly being torn apart by the sea, and the passengers were left bewildered and terrified in the cold, stormy dark.[2]

It was now time for the crew to man their lifeboat stations and to help the passengers board the boats in an orderly fashion, but that did not happen because neither the crew nor the passengers had been trained or, in fact, had been ordered to do so. Instead, because the captain had previously ordered the boats lowered to the rails of the ship, many of the passengers took that for permission to board the boats and cast off from the ship prematurely, often without the help of the crew.

This was a terrible mistake. Lifeboats, especially as designed and configured in the early 1900s, are finicky and demanding vessels, meant to be lowered in stages by professionals and manned by experienced sailors. In this case, though, people simply piled into the boats, many of them using knives to slash at the ropes that held the boat aloft. The result was that in almost all cases the boats were left with one end hanging significantly lower than the other rather than being evenly suspended. Once that occurs, it's very unlikely that the boat can be launched safely, and most of *Valencia*'s lifeboats were not. Instead they fell into the sea, dumping most or all of their occupants into the cold water, where they quickly drowned or succumbed to hypothermia. Most of those who survived long enough to approach the shore were beaten to death on the rocks by the savage waves.

Valencia pictured as she struck the reef during the storm. At this point, she is still bow-on to the shore.
ILLUSTRATION BY MOLLY DUMAS. USED WITH PERMISSION.

The deck was now crowded with terrified passengers. A mysterious figure in oilskins and sea boots, whom Bunker took to be a member of the crew, grabbed one of Bunker's children and led Bunker, his wife, and his other child over to lifeboat No. 6, which hung on the port side of the ship toward the stern.[3] (Recall that, the ship having pivoted, the bow was now pointing out to sea.) Having entered the boat, along with several others, four men stationed at the davits lowered the boat to the sea; this is one of the few instances in those by now early morning hours in which a lifeboat was lowered in a relatively, though not perfectly, safe manner. Bunker and his family were, for the moment, safe in lifeboat No. 6.

Chapter 15

Hostile Sea, Hostile Shore

Valencia struck Walla Walla Reef about midway between Cape Beale and Carmanah Light and just east of Pachena Point on the west coast of Vancouver Island, British Columbia, at—as near as can be told—11:50 p.m., the night of January 22.[1] She was nowhere near any inhabited towns, although a couple of First Nations settlements, including the village of Clo-oose, were not too far away. As described in the official report from the investigatory commission established by President Theodore Roosevelt, "The coast line where the *Valencia* lay is substantially a continuous rock cliff, rising almost sheer from the water, about 100 feet in height, covered with trees and beaten by a very heavy surf."[2] In other words, the ship had struck the rocks at just about the worst possible place for a shipwreck: The nearby shore was close enough to be tempting, but the waves and the rocks made it almost suicidal to attempt a landing. At the same time, rescue from seaward was unlikely due to the rocks, the rough surf, and the shallow water. *Valencia* was stuck between two equally deadly alternatives, at a place described by one source as "notorious for violent currents and unpredictable weather that have shredded steel ships down to piles of rust."[3] She was stranded between a hostile sea and an equally deadly shore, and help would come from neither quarter.

It's noteworthy that, prior to *Valencia* striking the rocks, the crew was never warned that the ship was approaching a reef. The lookout, posted on the bow, never called out, in spite of the fact that this was exactly the reason he had been stationed there. There are three possible reasons for this. First, the bow of *Valencia* was extraordinarily long; a crewman

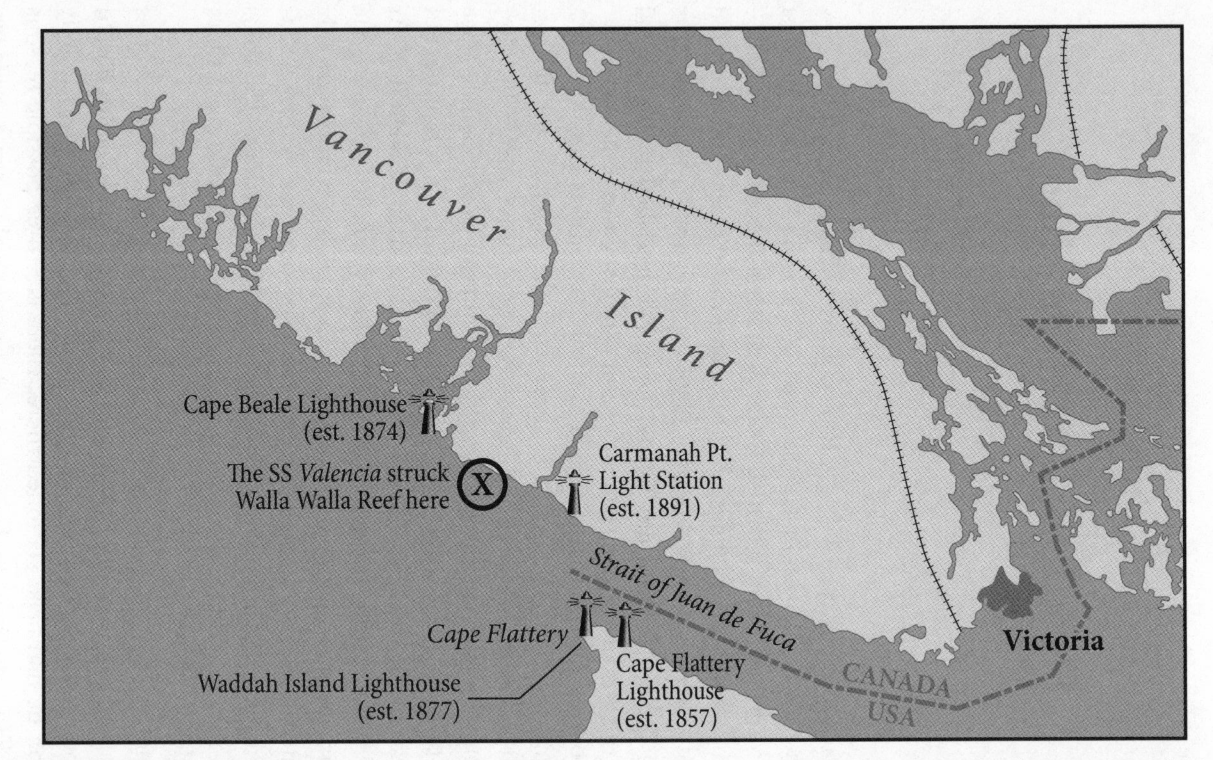

This map shows the point on the Vancouver Island coastline where *Valencia* struck the rocks. Note that there were two lighthouses on the coast, but neither was near enough to warn the ship as she approached the rocks in the middle of the night during a storm.

positioned there was quite a distance from the bridge where the officers worked the helm, far enough for a cry to remain unheard, especially over the noise of the storm.

Second, it was very dark, extremely foggy, and quite stormy. It was difficult to see anything directly in front of you; successfully peering through the rain and windswept sea foam to spy dark rocks against a dark background would have been a superhuman accomplishment.

Neither of those issues was Captain Johnson's fault. He did not design the ship, and he certainly had no control over the weather. However, he *was* ultimately held responsible for the fact that the lookout may simply have been *too tired to do a good job*. According to the testimony of survivors, including Second Mate Pete Peterson, the typical watch period of a lookout on the *Valencia* was somewhere between six and nine hours. In the minds of experts who testified after the wreck, that was entirely too long. A lookout could not have been expected, especially in the dark and in bad weather, to stand a watch of more than two hours or so; keeping a man at his post for six or more hours was almost certain to result in an exhausted—perhaps even sleeping—crewman, unable to keep a watchful eye out for hazards. Thus, the commission expressed its "strong disapproval" of the watch schedule, noting that "it is a matter of common knowledge that such long continued attention will dull the faculties so that a man would be much less quick and attentive than a fresh man."[4]

Frank Bunker was one of the lucky few to escape the ship on a lifeboat, but few on that boat survived. Even with men at the davits above carefully lowering the boat to the water, the lifeboat smashed against the side of the ship several times due to the rough weather and the surging sea. Also, even with a certain amount of care taken, lifeboat No. 6 was not really lowered horizontally; instead first the bow and then the stern were lowered alternately, and the occupants had a difficult time hanging on and staying in the boat. It's no wonder that most of the other lifeboats, lowered in an even more reckless fashion, were destroyed or simply dumped into the water.

When lifeboat No. 6 reached the water, the real dangers had just begun. Several men, including Bunker, used oars to shove the boat away from the *Valencia*, so that the smaller vessel would not be crushed against

This starboard navigation lamp is one of the few extant artifacts from the *Valencia* sinking.

PHOTO BY LESLEY SCHER. USED WITH PERMISSION. COURTESY OF HEATHER FEENEY, MARITIME MUSEUM OF BRITISH COLUMBIA.

or capsized by the larger. Still, as she drifted away from the ship, the lifeboat was almost immediately capsized by a wave that broke over the boat, tumbling people into the sea. One of those people was Bunker, who, upon surfacing, swam toward the overturned lifeboat. Another wave turned the boat right-side up, though it was now filled with water. Climbing into the boat, Bunker was astonished to find his wife, Isabelle, still sitting in the seat she had occupied previously. The Bunkers' two-year-old son was still in the water, but Bunker retrieved him by grabbing on to the boy's life vest. The child was unconscious, but Bunker was able to revive him, and the boy, terrified and cold, began to cry weakly.

By this time, there were only four occupants in the boat: Bunker, his wife, his son, and Frank Richley, a fireman from the *Valencia*. Bunker's six-year-old daughter, Dorothy, had vanished beneath the waves, but there was no time to grieve her loss. A combination of wave action and paddling brought the boat almost to the shore, but then a wave pushed the boat violently onto the rocks. Bunker lost his grip on his wife and son; the two were washed away and never seen again. Bunker was then tossed against the rocks, finally grabbing hold and scrambling to the pinnacle of an outcropping. He crawled on his hands and knees up the rocks along the shore, looking for his family, but they were not to be found. Bunker was alone, as far as he knew, cold and wet, his clothes and skin shredded by the sharp rocks.

Climbing to the foot of the bluff that stretched above the rocky shore, Bunker heard voices calling. He encountered a group of eight men, including Frank Ritchley, the fireman from the ship, who had been in lifeboat No. 6 with him.[5] The nine of them huddled together at the bottom of the one-hundred-foot cliffs. One had matches, so they attempted to start a fire using a life preserver they had soaked with oil found in the overturned and beached lifeboat, but, not surprisingly, the matches were too wet to light. It was a long, cold night spent waiting for the sun to rise, all of them thinking about family and friends they had lost and worried about what the morning would bring.

Chapter 16
Safety Measures

Much has been made, in this book and others, of the lack of lifesaving infrastructure in and around British Columbia at the time of the *Valencia* incident. There were few lighthouses or lightships in the area, largely due to the disinterest of governmental bodies—Canadian *and* American—in charge of such undertakings in the area. In spite of the fact that the Pacific Northwest hosted so many violent storms that the area was, even then, known as "the graveyard of the Pacific," authorities were reluctant to devote time, money, and resources to what must have seemed to them an underpopulated backwater. One assumes that such resources were devoted instead to more sophisticated—and more densely populated—coastal cities, such as Boston; New York; Halifax; and Moncton, New Brunswick. Similarly, there was also a dearth of lifesaving stations in the area, again in spite of the number of wrecks reported.

As retired Canadian Coast Guardsman Clay Evans notes, it was difficult getting infrastructure built on the island. "We've always felt like the forgotten stepchild, and there's still an element of that today. Because our nations, both the United States and Canada, developed from east to west, there was, particularly back in 1906, always more infrastructure back there [that is, on the East Coast] than there was here."[1]

But in addition to all that, we must consider the lifesaving gear carried on *Valencia* herself. We've already noted the lifeboats, of which there were six, hung, three on each side, on davits, plus a "working boat" that could function as a lifeboat. The seven boats could carry, altogether,

181 people. There were also three life rafts that together could carry an additional forty-four people.

Valencia also carried 368 life preservers (most made of tule, a buoyant reed about which we'll learn more shortly, and the others of cork) and a Lyle firing gun (a small, specialized cannon, really) with line; this device was used to fire a line from the ship to the shore or vice versa, where it could be picked up and attached to a larger rope or a bosun's chair or breeches buoy meant to ferry people from a stranded or sinking boat to the shore. The ship was, as noted in the investigators' report, appropriately "equipped as the law and regulations required."[2] However, some charged that those rules and regulations were either too lax or, worse yet, undercut by political machinations that allowed inferior materials to be used in such things as rafts and life preservers.

The "inferior materials" in question were the tule life preservers. Some of the survivors claimed that life preservers made of tule were worse than useless: they became waterlogged very quickly, said some, after which point they were simply a sodden mess that was more likely to drown a person than support his weight.

The government listened to the complaints and even tested a batch of eight life preservers—immersing them in water for seventeen hours and hanging weights from them—and determined that they stayed buoyant for the appropriate period of time. In any case, the inspectors could find no official fault with their use, "as such life-preservers were allowed by law and had been properly passed by the inspector."[3] Even if they were flawed, the official report seemed to be saying, they had nonetheless passed inspection and any criticism should be leveled at the inspecting agencies, not the life preservers themselves or the ship's owners or operators. The investigators did recommend, however, that the Department of Commerce and Labor "continue its exhaustive series of tests as to the merits of the tule life-preserver, and collect information thereon from all possible sources and submit all the tests already made, and hereafter to be made, to the Board of Supervising Inspectors."[4] What became of the results of those tests is unknown. *Valencia* actually carried *more* life preservers than were required, but the question remains: Were the tule life preservers, whatever their number, adequate for their task?

Frank Bunker later charged that the life preservers on board *Valencia* were unserviceable, and that any previous inspections were made simply to ascertain that the ship carried the appropriate *number* of them rather than to determine their actual condition or efficacy. Shortly before *Valencia*'s last inspection, the US Navy's Rear Admiral Louis Kempff, superintendent of the Pacific Naval District, had conducted his own inspection of boats in the Tacoma area and had condemned most of the tule life preservers found on workboats in the area. Nonetheless, the same type of life preservers had been allowed on *Valencia*.

Industry opinion of tule as a life preserver material was mixed at best. When another Pacific Coast Steamship Company vessel, the *Santa Rosa*, sank in 1911, the officers on board specifically recommended that the passengers *not* use the provided tule life preservers. One source quotes an officer as saying, "They're filled with tule . . . and they won't float. We always buy our own cork jackets and keep them with us when we're working for Pacific Coast Steam."[5]

The most serious charges made may have been those leveled against some of the inspectors themselves. Inspector Bion B. Whitney, working for the US Marine Inspection Service, defended the life preservers, noting that his agency regularly had men visit the factories where the life preservers are made, rejecting those that didn't exhibit the required buoyancy.

Whitney, however, was criticized as lacking experience. At the time, the law required inspectors to have served at least five consecutive years as masters of ships and to have practical knowledge of ship construction. Whitney seemed to meet few of these requirements: according to one source, he had been on only three voyages—logging a total of six months at sea—and may have been the ship's master on none of them.[6] There was also some question as to whether Whitney and his partner might be inclined to "give the Pacific Coast company the best of it," in other words, that he might wish to give the company a break for political reasons.[7]

In the end, nothing was ever proven as far as the inspectors' lack of objectivity, or relating to the usability of the life jackets themselves, other than we note that many sailors held them in low regard, preferring other materials.

At the time of the *Valencia* incident, Rear Admiral Louis Kempff was in charge of the Pacific Naval District and conducted his own tests of the tule life preservers being used on boats in the Tacoma area.

Chapter 17

Chaos

When the *Valencia* struck the rocks, Captain Johnson's initial reaction was to shout, "My God! Where are we?!" This sounds like an odd thing to say, but it made sense, because he literally did not know the ship's location; after all, if she were where he had thought she was, she would not have hit a reef. He must have been astounded that his ship had run aground out here in what he had thought was "the middle of the ocean." In reality, he must have had *some* inkling that they were off course enough to have been almost anywhere, especially with the fog obscuring everyone's vision, because the ship's forward speed had been reduced to dead slow.

As it turns out, where they were was easy enough: they were stuck on a reef only yards offshore of the west coast of Vancouver Island. How they were to get off the reef, if that was at all possible, was the question. Johnson then pivoted the ship such that her stern was only yards from the cliffs. This placed salvation, in the form of land, temptingly close; if it had been daylight, the passengers and crew could easily have seen the rocks that formed a sort of beach, and the cliffs that loomed overhead.

But that salvation was an illusion. There was no safe way to get the people left on the ship to the shore while the storm raged, even though it was so near. In the meantime, the ship itself was becoming more dangerous and less stable as the minutes passed. The crew checked the bilge and noted that water was rising there at the rate of about one foot per minute. As the water reached the ship's generators, the lights failed, and huge waves continued to batter the stranded vessel.

It was time to lower the lifeboats, this time officially and completely, but first the crew had to convince the remaining passengers, some of them reluctant to leave the perceived safety of the ship. The captain, of course, knew that any supposed safety was an illusion; the passengers—and especially a group of women who were insistent about staying with the ship—would be much safer on the lifeboats. The ship, he knew, was being torn apart by the sea and the storm.

This is the first appearance of an issue that we'll encounter repeatedly over the next several hours, as the sea whittled away at the ship: some passengers, especially the women, were reluctant to board the lifeboats, at first because the ship seemed fairly stable—or at least less threatening than the wild, tumultuous ocean—and later because they thought that the ships they could see in the distance would soon arrive to rescue them.

Both assumptions were incorrect. *Valencia* was unsafe and becoming more so as the minutes passed. When ships finally did appear on the horizon to attempt a rescue, many of the women would again insist on staying aboard to wait for help to arrive. But the ships, as we will see, would never approach closely enough to mount a rescue attempt, and the women who insisted on staying with the ship thereby condemned themselves and their children to death.

As it happens, in 1906, few women knew how to swim, so that may have played a role in their decision. (And really, not that many men were competent swimmers at the time; ironically, even many *sailors* did not know how to swim.) There were many reasons for this. According to the social norms of the day, women were expected to be modest and demure; swimming was seen as a physically demanding—and somewhat revealing—activity that was not, according to some, appropriate for women and girls.[1] As a result, not many women knew how to swim in the early 1900s.[2] All of this would change over the next few decades, but in 1906, it was quite likely that few of the female passengers were good swimmers. They would naturally be reluctant to enter the water, especially in the middle of the night during a storm, and certainly when the ship itself felt safer than the lifeboats, many of which had capsized or been crushed before their eyes.

This swimsuit, although it covers just about all of actor and competitive swimmer Annette Kellerman's body, was considered quite daring, compared to the pantaloons usually worn by ladies at the beach during the early 1900s.

Second Officer Peter Peterson was frustrated by this reluctance to leave the stricken ship. He testified that, even after convincing five women to get into a lifeboat, two of them demurred, saying that they'd rather "stay by the ship."[3] This reluctance to board the lifeboats, and the three life rafts the ship carried, even when urged to do so by crewmembers, would result in the deaths of every woman and child that had not already died in the sea or on the rocks. (Then again, every woman who did enter a lifeboat *also* died, so who's to say which choice was wiser?)

After she struck the rocks, chaos reigned on *Valencia*. Most of the lifeboats were inundated by passengers who did not wait for instructions from the crew or for orders to board; they clambered aboard the boats that, instead of being lowered correctly, were simply dropped into the sea. Few survived, and most of those who did were soon drowned or pounded to death on the rocks lining the shore. Children, abandoned in the tumult, cried for their mothers, many of whom had been lost in the lifeboats or swept overboard by the towering, crushing waves. One mother handed her child to her husband in a lifeboat, but the child was immediately swept away and fell into the sea; moments later, the husband drowned when the lifeboat overturned. A miner who had struck it rich in Alaska offered all of his money, $1,800 in gold, to anyone who could take him ashore. There were no takers, and the gold simply disappeared, trampled underfoot by the terrified crowd on deck. Money, even a small fortune such as that offered by the miner, was not worth dying for.

The people on board, especially the passengers, had to have been terrified. It was dark and cold, and the ship was being torn apart even as they sought shelter on deck or in the rigging. The ship's cargo—vegetables, wine, and more—was strewn about the deck, which was awash in cold water and broken spars, lengths of line, and tangles of rigging, all of it making footing treacherous; tripping could mean falling headfirst over the side or sustaining a serious injury. The deck of the SS *Valencia* was not a safe place to be, and it was becoming more unsafe by the minute.

The terrible irony—among many terrible ironies—was that if the boats had been launched correctly, and especially if people had waited and launched them the following day, many more passengers and crew could have been saved. Of the three lifeboats launched early on, only nine people—all men—survived. In fact, one boat that was left after the debacle *did* launch on Tuesday morning, making it to shore with little difficulty, the storm having calmed somewhat and visibility being much better in the daylight. The seven occupants of this boat would make up the so-called McCarthy party, which traversed the trail above the bluff to the east, looking for help.

One of the nine who survived the early lifeboat launchings was Frank Bunker. He and his group, which would become known as the Bunker party, struggled and scraped their way to the top of the bluff. They decided that their best course of action was to head west on a narrow, poorly maintained trail at the top of the bluff, in order to seek help. They turned left, a decision that may have doomed those on the ship who were still awaiting rescue.[4]

Chapter 18

The Bunker Party

It was now early Tuesday morning, January 23. The situation on board *Valencia* was rapidly deteriorating as the ship disintegrated under the relentless attack of the wind and the waves, and things on land were not going much better.

The nine survivors of two different lifeboats included Frank Bunker; firemen Frank Richley and George Billikos; first-class passengers R. Brown and Charles Samuel; second-class passengers T. J. Campbell, Michael Howe, and Yosuki Hosoda; and naval cadet J. Willits.[1] (Some sources say that Bunker was booked in a first-class cabin, while others—including the official investigating committee—say that his accommodations were second class.)[2] Together, they milled about, stamping their feet and waving their arms to help keep off the chill. Rain and cold wind assaulted them as they waited for daylight, when they hoped to find a way up the bluff to where there might be some sort of trail leading to civilization.

As dawn neared, Bunker headed up the bluff alone, resolved to see if it was accessible. Determining that it was, he returned to the other eight and led them up to the top, where they found the barest of trails—a much-used game track that had been somewhat widened for use as a trail for servicing the telephone and telegraph lines that ran to the east and west.[3] The group turned to head west, but, according to Bunker, their flight was halted when one of the men informed Bunker that he had seen another man—possibly insane, certainly raving—down on the rocks below.

At this point, Bunker begins to paint himself as a bit of a hero. (And in the coming hours, something more than a bit.) Telling the investigating committee that he couldn't possibly desert a man down on the rocks until everything had been done to ensure his welfare, he says that he "climbed back down to the beach [and found a man] whose face was smashed in by the rocks; he was insane." Keep in mind that Bunker, while no fool, was not a psychiatrist; how he determined, and determined so quickly, that the man was insane is never explained. The man might have been raving, or he may simply have been freezing, injured, and understandably terrified. Bunker says that the man kept struggling to get back into the water, but that Bunker restrained him and laid him down on two life preservers so that he would be out of the reach of the waves, should he recover his senses. Bunker then returned to the top of the bluff; the "raving" man is never mentioned again.

At that point, the Bunker party could have headed in one of three directions. Bunker's initial impulse was to head inland, but he opted against that. This was a wise decision; from their current location, they could have traveled many miles inland and never seen a soul, let alone encountered a town or even a settlement of any size.

Gathering the men together, Bunker pointed out that they needed to make a firm decision about the proposed direction of travel; wandering aimlessly or constantly second-guessing one another was certainly not going to be helpful. They decided to go west, as was Bunker's preference. We can see, and it's important to keep in mind, that most of these men were not experienced outdoorsmen. The sailors from the ship were certainly hardy men, but neither they nor the passengers were necessarily experienced overland hikers, adept at finding their way and surviving in the wild.

And this *was* a wild land. At the time—and even now, to a great extent—there is very little in the way of habitation, civilization, or infrastructure in the area where the ship struck the rocks. The survivors had no way of knowing, of course, but they were miles from any sort of settlement, and the closest outpost of any kind was the lighthouse at Carmanah, roughly fifteen miles or so southeast of the wreck. Even today, the only way to view the wreck site is by boat, keeping well clear of the rocks and reefs that destroyed *Valencia*, or to hike several miles overland to a remote stretch of coastline, with little in the way of comforts or amenities along the way. The trail there, now maintained by the Canadian government as part of the Pacific Rim National Park Reserve, is today almost as difficult an overland route as it was back when the *Valencia* survivors walked it. (Parks Canada notes that hiking the entire trail can take several days and covers "rugged, uneven ground," and that one will likely encounter muddy trails, boulders, and rocky shorelines. See chapter 29 for more about the West Coast Trail.)[4]

The Carmanah Point lighthouse, seen here in a 1994 photo, has been active since 1891 and was located some fifteen miles southeast of the spot where *Valencia* struck the rocks.

The group was at the time just south of Pachena Point, about eight miles southeast of the village of Bamfield, British Columbia, as the crow flies. Of course, the meandering route one would have to hike to cover those eight miles would end up being a journey of much longer than eight miles because a hiker would have to traverse multiple rivers and either sail across or walk around Pachena Bay in order to reach Bamfield, which, even now, boasts a population of fewer than two hundred.

The party was somewhat northwest of the First Nations village of Clo-oose, but it's important to note that the group did not at first *realize* they were in British Columbia at all. Having so far seen nothing to indicate otherwise, they thought that they were just south of Cape Flattery, on the American side.[5] Knowing that there was a lighthouse at Cape Flattery, they assumed, correctly, that there must then be a telegraph line leading to the light, and perhaps stations along the way. As it happens, they would indeed encounter a telegraph line that they could use to communicate with potential rescuers, but it was not the line that ran to Cape Flattery, which was in an entirely different country, twenty-five or thirty miles away, across the Strait of Juan de Fuca, itself about ten miles or so across.

Of course, there had been another option to begin with, and it is one the group should have considered more seriously. The survivors, rather than head east or west, or indeed, rather than go anywhere at all, could have stayed more or less where they were, simply making their way back to a site above the ship. Firemen Ritchey and Billikos, the only ones in the group with any real seafaring experience, objected to moving out, preferring to stay put, in case someone on board *Valencia* managed to get a line to the shore. If they left, the two men reasoned, there would be no one to take up such a line.[6]

With Billikos and Richley outvoted, the men hiked through ankle-deep snow, some of them without shoes, until about 1:00 in the afternoon, at which point they left the trail and descended a hill down onto the beach. Finding a cabin there—which they correctly assumed to be a lineman's hut—the cold, exhausted men built a fire in the stove and Bunker managed to tap into the telephone line and speak to Mrs. Logan, a lineman's wife, at Clo-oose, announcing the wreck and asking for help.

Mrs. Logan and her husband began spreading the word, and Mr. Logan contacted *Queen*, a liner due to arrive at Victoria, to be dispatched for rescue work.

Help was on the way, it seemed. Bunker's heroics, as he describes them, had managed to get the word out, informing the world of the *Valencia*'s plight. In the end, though, help would not be forthcoming and Bunker's group would be castigated—rightly or wrongly—for having left their shipmates to die.

Chapter 19

"It Was a Good Plan"

On the ship, Captain Johnson and his crew were taking what steps they could to save the souls on board, firing rocket after rocket to attract the attention of both passing ships and the survivors that they hoped had made it to shore. One surviving sailor testified that Johnson had blown off two of his fingers lighting rockets that had malfunctioned, possibly due to wet gunpowder, but he remained in command, giving orders and assisting as best he could.

Johnson's shrewdest move, made as soon as he knew that some survivors may have reached the shore and that another boat was about to leave, was to ready the Lyle gun to fire a light line from the ship to the shore. The idea was that the forward part of the line would be picked up by those on shore, while the "bitter end," that is, the end of the line remaining on the vessel, would be tied to a much heavier rope that would be hauled to the beach by the men on shore.[1] At that point, various devices—such as a breeches buoy or, if need be, a bosun's chair or even a simple bight of line—could be used to transfer passengers and crew from the ship to the shore.

It was a good plan, and it had been used effectively many, many times. The Lyle gun was a proven lifesaver, a short-barreled cannon that fired a projectile to which had been affixed a strong but thin line, and it had been used extensively since its invention in 1878 by a career Army ordinance officer named David A. Lyle.[2] Interestingly enough, Lyle had at that time been seconded to the US Life-Saving Service, then under the command of Sumner Kimball, for the express purpose of improving the design of

The Lyle gun, named after Army officer David A. Lyle, fired a small-diameter line from the ship to the shore—or vice versa. After connecting that line to a thicker line, one could haul people and supplies from the ship to the shore, as long as there was someone ashore to secure the line fired from the ship.

the gun, which had been first used in the early 1800s, but which proved to be inefficient and relatively ineffective at first.

Kimball himself was a lawyer who in 1871 had been appointed to overhaul the Revenue Marine Division and made it his life's work to improve its lifesaving effectiveness. Kimball modernized the organization, professionalizing its staff (previously the organization had depended mostly on volunteers) and establishing a training school that eventually became the US Coast Guard Academy. In the words of historian Dennis Noble, "Kimball was unquestionably the driving force behind the United States' possessing a first-class lifesaving organization. Much of the present-day Coast Guard's highly regarded reputation as a humanitarian organization is the result of his organizational skills and management abilities."[3] In the annals of US lifesaving heroes, no one stands taller or shines brighter than Sumner Kimball. It was fitting—perhaps inevitable—that David Lyle worked under Kimball to improve the device that *might* just save everyone left aboard *Valencia*.[4]

Of course, the effectiveness of the Lyle gun relied on the line being placed in the right spot (in this case, the top of the bluff would be best) and on someone being there to grab and secure the end of the line. The

Sumner Kimball revitalized and professionalized the US Life-Saving Service, leading directly to the development of the US Coast Guard.

fate of the *Valencia* survivors would hinge on the people at both ends of the line being present and able to do their jobs.

In any case, there weren't a lot of other options for the remaining passengers and crew to consider. Attempting to swim to shore was almost suicidal—and keep in mind that it's unlikely that everyone could swim, in any case. At the same time, the ship was being torn apart by the storm and the waves. The people remaining on *Valencia* were trapped between two hostile forces: the bitterly cold sea—which would soon pound the ship to pieces and would quickly suck the life out of anyone who remained in the water for any extended period of time—and the shore itself, which was deadly because of the sharp rocks that guarded it and which would batter senseless those who attempted to climb them.

Boatswain Tim McCarthy had been asleep in his bunk when the ship struck the reef. Arising hurriedly, he went on deck, still in his nightclothes. Realizing the scope of the unfolding disaster, he returned to his bunk and put on clothes. Back on deck, in the dark and confusion, McCarthy couldn't tell passengers from crew, but at one point an officer told him to clear away the boats, that is, to swing them out from the davits and begin lowering them to the saloon rail and keep them there. However, seeing the small crafts being readied, the passengers rushed the boats, piling into them before they were ready to be safely boarded and released. According to McCarthy, only one of the boats, lifeboat No. 2, got away without incident. A few minutes later, the No. 1 lifeboat smashed to pieces against the hull, dumping everyone into the water; McCarthy saw one man climb back aboard *Valencia*, using some ropes and ladders that had been swung over the side, but he couldn't tell if anyone else had made it back on board.

McCarthy noticed one of the ship's firemen (or a fireman's messboy, testimony varies) on the No. 6 boat; that would be the boat on which Frank Bunker and others escaped the doomed ship. One of the three rafts also floated nearby, with about eighteen or nineteen people on it.[5]

By the time McCarthy spoke with Captain Johnson about mounting an attempt at a shore landing, the social hall and part of the saloon were the only dry places left on the ship. Many of the women and children had climbed into the rigging, some having bound themselves to the mast and spars with belts, ropes, and strips of cloth in order to keep from falling into the sea.

Captain Johnson had faith that perhaps McCarthy's party—or the Bunker party, which had left earlier—would be able to reach the shore, climb the bluff, and then retrieve a fired line. If so, they were saved: the line, once found and then attached to a stout rope, could be used to haul the remaining passengers and crew to safety.

On the *Valencia*, Tim McCarthy gathered volunteers to take the one remaining lifeboat to shore, while the rest of the crew manned the Lyle gun. Neitzel says there were six total in the McCarthy party, while the official commission report says there was "a crew of seven."[6] The latter number is reinforced by other sources, but one of the ones that names the other six men in the party—Thomas Shields, John Marks, William Gosling, Tom Lampson, Charles Brown, and John Montgomery—also points out that Montgomery did not, in the end, accompany the group. He was busy helping launch the boat, manning the rope and pulley used to lower it into the water. Afterward, he attempted to jump into the lifeboat but instead plunged into the sea and drowned in the turbulent water.[7] This would seem to confirm Neitzel's count.

McCarthy was a skilled seaman with fifteen years' experience; he had gained his position as bosun, or deck boss—the most senior of deckhands—after having many times proved both his abilities and his courage. If anyone had a chance of landing safely and securing the line fired from the Lyle gun, it was McCarthy.

But the ship was quickly being torn apart, with water now over the lower decks, the pilot house, and the bridge. Food was scarce because most of the stores were now underwater. There wasn't much time left to act. The combination of the Lyle gun and the people on shore was their last, best hope.

Chapter 20

"The Commission Does Not Desire to Attribute Blame . . ."

By the time the Bunker party had reached the shore, searched for a pathway up the bluff, and climbed to the top, they were somewhat west of the wreck itself. If they had turned east, they would have found themselves atop the bluff in a spot overlooking the wreck. Had they then stayed there, they could have—almost certainly would have—been able to spot and secure a line fired from the ship, which was well within range of the Lyle gun: the gun could fire its projectile, with a line attached, close to seven hundred yards, and *Valencia* was only a few hundred feet from shore.

In the words of the federal commission's report, the group "would unquestionably have been there to receive the line which, as a matter of fact, was later fired there from the vessel; they would have had 9 men to pull ashore the line and attached rope and to make it fast."[1]

Still, the investigators were careful not to judge Bunker and the rest of the party too harshly: "The Commission does not desire to attribute blame to this shore party for its failure to grasp this opportunity. They had been beaten through the surf, some of them considerably injured, had spent the night in great discomfort, without food or shelter, and were none of them in a condition to coolly estimate the chances and to consider expedients which occur to persons in normal conditions." The investigating committee was at pains to allow the men as much leeway as possible and to refrain from some of the finger-pointing that had been going on in the wake of the disaster, "inasmuch as throughout the history of this disaster blame has perhaps at times been hastily imputed to various individuals and officials for failure to think of exactly the right thing at the right time."[2]

Hindsight is all well and good, but under those conditions, the committee seemed to be asking: How many of us could have done better—or even as well?

Having dismissed the objections of Billikos and Richley, the nine men of the Bunker party headed west, hoping to run across a town or other evidence of civilization. They would soon encounter the lineman's hut mentioned earlier and tap into the telephone line to make contact with the lighthouse keeper at Cape Beale. This would indeed help spread the word of the disaster, but it would, in the end, do nothing to help the people left on the ship; they would soon die, as the sea slowly tore apart their vessel and no one came to rescue them.

Some hours later, and several miles northeast of the wreck, McCarthy and the occupants of lifeboat No. 5 were fighting the sea and the rocks. They were cold, tired, and sore, but just after 1:00 p.m. on Tuesday afternoon, they managed to make a safe landing on shore. They continued heading northeast on foot, finally spotting a telegraph line, which they followed to a cabin.

There the men made a startling discovery. Like the members of the Bunker party, they had thought they were on or near Cape Flattery, on the US side of the strait of Juan de Fuca. But there, next to the trail they had followed to the cabin, was a sign: "Three Miles to Cape Beale." Cape Beale, they knew, was not in the United States at all; they were in Canada, on the west coast of Vancouver Island, British Columbia, just south of Bamfield. They knew of Cape Beale because the mariners among them knew their lighthouses, and there was a lighthouse at Cape Beale, at the southern tip of Barkley Sound, near Pachena Bay. (The lighthouse, established in 1874, is still active today, though it has since been redesigned and rebuilt.)

Realizing that help was, in theory, only three miles away, the McCarthy party decided that it was wiser to head for the lighthouse rather than

The Cape Beale light, shown here in a 2006 photo, was erected in 1874.
KENWALKER

try to head back to the site of the wreck. (They would later describe the forest as "impenetrable," in any case, which it probably was in 1906. Hacking their way back through the dense underbrush might've taken several hours, and time was of the essence.[3]) Following the trail to the lighthouse was much faster and easier, but it was an unfortunate choice; if they had managed to make it back to the wreck, they might have been able to pick up a line fired by the Lyle gun, thus saving everyone left on board the ship. Nonetheless, it was an understandable decision: if they reached the lighthouse, they could get in contact with rescue vessels and perhaps others to mount an overland rescue mission. (Which in fact is what happened, though in the end that mission didn't do any good.)

The Bunker party also followed the telegraph and telephone lines that were strung together through the trees for several miles, finally descending the trail to the beach, only to encounter a river. This was the Darling River, which runs to the north inland of the coastline, skirting several First Nations reservations.

When Bunker and his group reached the river, it was running high, swollen with runoff from recent storms, but they could see a cabin on the opposite shore, with the telegraph/telephone wire running right to the cabin; they had no choice but to try to cross the river. Bunker went first, after having tied a rope around his waist. He made it across, secured the rope, and the other men followed, using the rope to guide them.

In the cabin—really a simple lineman's hut—the group found supplies, including a stove, some bacon, and some lard. Most importantly, they found a receiver made for both telegraphy and telephony. Bunker used it to contact lineman David Logan and his wife at their home in Clo-oose, explaining that he was part of a party from the wrecked *Valencia*. Logan then forwarded that message to the lighthouse keeper at Carmanah Point and arranged for two men, one of whom was Phil Daykin, the lighthouse keeper's son, to meet Logan to begin a search for the wreck.

Shortly after that, Bunker was able to send a message to Minnie Paterson, wife of the keeper at the Cape Beale lighthouse. Minnie Paterson deserves her own chapter in this tale, for she is one of the true heroes—we might say, one of the *few* real heroes—of the story. For seventy consecutive hours, she stayed at the lighthouse, answering calls, sending out word of the disaster, and clothing and feeding those who showed up.

This was nothing new for Paterson; she was well known in the area for her strength, her dedication, and her perseverance. In fact, later that year, Minnie would make her famous run from Cape Beale to Bamfield to summon help for the crew of the *Coloma*, foundering off of Cape Beale. With the telegraph line down, as it often was, Paterson walked for four hours through mud and water until reaching help in Bamfield. She then turned around and *walked back home*, so that she could tend to her children.[4] Paterson, who would die in 1911, was nothing if not a hardy soul.

Soon after Paterson received Bunker's somewhat garbled message, McCarthy and the other five members of his group showed up, having

Lineman David Logan and friend. Logan was part of the three-person search party that sought—and found—the wrecked ship, but too late to do any good.

Minnie Paterson was one of the few heroes of the *Valencia* incident.
ILLUSTRATION BY MOLLY DUMAS. USED WITH PERMISSION.

trudged the three miles to the Cape Beale lighthouse. Paterson greeted them with an apology, noting that she was sorry she could not contact them. This confused McCarthy, until he realized that Bunker's group had already called in to report the wreck.

McCarthy asked Paterson to contact Victoria or Seattle for assistance, which she did, first telephoning the men in the cabin near the Darling River (Bunker answered the call) and then telegraphing the authorities in Victoria and beyond.

Assistance was now on the way, but would it come too late? When it arrived, would there be anyone left alive to save?

Chapter 21

The End of Hope

On board *Valencia*, things were going downhill, and the situation would soon get worse. Captain Johnson had hopes that McCarthy and his crew would reach land, but he was ready to consider just about any other option. Fireman John Segalos (or Joe Cigalos, according to some sources) volunteered to try to swim a line ashore. He tied one end of the rope around his waist and dove into the turbulent, freezing water.

He didn't get far. About halfway across, the line became entangled, so he cut it and headed back to the ship. Pulled aboard, those who had watched the attempt gave him dry clothes and offered him whiskey, the latter of which he is said by some sources to have refused, although Neitzel has him testifying that he took it.[1] Either way, it was a valiant effort, but the odds against Segalos were overwhelming; in the frigid, frothing sea, it was impossible for a man to swim a line to shore. The passengers and crew were out of options. It was up to the two shore parties and those they had summoned for help.

Requests for assistance had by now reached Victoria, where Captain James Gaudin, of the Canadian Department of Marine and Fisheries, received a series of telegrams informing him of the wreck. Gaudin began to spread the word, and soon three ships set out from Victoria to attempt to aid in the rescue. Two of them were *Czar*, an oceangoing tugboat, and *Queen*, a steamship that Captain Johnson had in fact captained several times before, but which was currently under the command of Captain N. E. Cousins. The last vessel was *Salvor*, a wrecker or a salvage steamer—that is, a ship that specialized in retrieving salvaged goods from shipwrecked vessels.

The weather was still bad, so it took several hours for the ships to meet up at the Carmanah Point lighthouse, near the mouth of the Strait of San Juan de Fuca. They set out at first light for the wreck site, and at 8:30 the next morning, mariners on board *Queen* spied *Valencia* on the reef. The sailors saw smoke on board from fires, most likely lit both to warm the passengers and to attract the attention of nearby ships. From the deck of *Queen*, sailors reported seeing figures on board *Valencia*, many of whom had lashed themselves to the rigging. The pilot on board the *Queen*, Herbert F. Beecher, also reported to the Associated Press that, although *Queen* got no closer than about half a mile from the wreck, he could see—with the aid of binoculars—passengers lashed to the rigging of the *Valencia*. However, Beecher also testified that it was unsafe to launch any boats, as "the sea would have smashed them before they could get the oars out."[2]

But there was an issue. *Queen*, being a large ship, was unable to get close to *Valencia*, as the latter vessel was stuck on a shallow reef. *Queen* was 331 feet in length, with a beam of some thirty-nine feet. She drew twenty-one feet and was therefore unable to enter shallow water, so her captain backed *Queen* off and let *Czar* approach the wreck.[3] The sailors on *Czar*, though, said that they saw no signs of life, in spite of Cousins's earlier report that they had seen smoke and people aboard *Valencia*. Having determined that there was no one left to save, *Czar* and *Salvor* both left the area. The cold, wet, frightened people on *Valencia* had to have been horrified, watching the two vessels sail away; surely, they must have thought that the two ships were coming to rescue them.

Queen remained on station, though, her captain believing that there *were* in fact survivors aboard *Valencia*; the captain and crew began discussing rescue possibilities. The ship had lifeboats, of course, but with swells running ten to fifteen feet or more, Cousins worried that the lifeboats would not make it to *Valencia*—or, if they did manage to reach the stricken ship, that they would not make it back. Taking that kind of risk might mean losing both his lifeboats and the crewmembers manning those boats—not to mention the *Valencia* passengers they were

Capt. James Gaudin of the Canadian Department of Marine and Fisheries

attempting to save. Note, though, that many were of the opinion that lifeboats could have been launched safely. A *Seattle Times* headline of January 29 notes that Second Officer Peter Peterson was one of the ones who believed that it would have been, if not actually *safe* to launch rescue boats, at least worth the risk.[4]

It is at about this time that, apparently due to what the investigating commission called a "mistaken understanding," rumors began to fly implying that a group of US Navy sailors aboard the *Queen* had volunteered to take a lifeboat to the wreck. Captain Cousins has been roundly castigated for supposedly refusing that offer. However, the commission found—and several of the Navy sailors aboard testified—that no such offer had been made; evidently the sailors shared Cousins's opinion that to risk such an attempt was foolhardy and would only result in more loss of life.[5]

Around that time, another steamship, *City of Topeka*, joined the group. On board *Topeka* was J. E. Pharo, an assistant manager of the ships' parent company, the Pacific Coast Steamship Company. Pharo, who was the senior company officer present, told Cousins to return to Victoria

The steamer *Queen* entering New York harbor

with *Queen*, which Cousins did, after which a dispute broke out: Captain Patterson of the *City of Topeka* argued that it was *he* who was in command of the vessel, and that Pharo was in no position to give such orders; he maintained that what he received from Pharo and others were merely "suggestions." It's important to keep in mind that it was Pharo who had ordered Cousins to take *Queen* to the wreck site in the first place, so one assumes that he would have the right to recall her.[6]

Regardless of who issued the order for *Queen* to return to Victoria, the federal investigators took issue with it. They agreed that Captain Patterson was the one in charge there, and that Pharo should never have ordered *Queen*'s return. The committee therefore noted that "the legal responsibility for this order rests upon Mr. Pharo and the moral responsibility upon Captain Patterson, and that both of them are highly censurable for having issued or sanctioned this order."[7] Of course, that after-the-fact censure did no one on board *Valencia* any good. In fact, it may be that Pharo was simply doing his perceived duty to the Pacific Coast Steamship Company, wanting the line's ship back in action and making money for the company as quickly as possible. And back in action she soon was: returning to Victoria, *Queen* immediately loaded her passengers and cargo and embarked on her scheduled voyage to San Francisco, leaving *Valencia* to her fate.

The *City* of *Topeka* steaming away from Port Townsend, Washington, in 1898 or 1899

FROM THE UNIVERSITY OF WASHINGTON COLLECTION

To Pharo's credit, he *had* attempted to dispatch tugboats to aid in the rescue, only to be told that none were available. He had been misinformed, though: as it happens, there were some tugs at Neah Bay, some fifty or more miles from Victoria, but the telegraph line between the two towns was out, so there was no way to contact them. Nonetheless, the investigating committee commented that Captain Patterson knew—or should have known—that tugs were available at Neah Bay, and noted, "Either Mr. Pharo or Captain Patterson should have directed the *Queen* to stop at Neah Bay for a tug on the way down."[8]

The entire *Valencia* incident abounds with tragic "should haves," "could haves," and "if onlies." In this case, one oceangoing tug brought down from Neah Bay, located only about forty miles from the wreck, near the mouth of the strait on the US side, may have saved many lives. If only the telegraph line weren't down, if only Pharo or Patterson had arranged to pick up a tug or two at Neah Bay, if only . . .

The *City of Topeka*, meanwhile, stayed on task, looking for *Valencia*. But with visibility poor, *Topeka* steamed right past the ship, never sighting her. In addition to there being poor visibility, it's interesting—and perhaps somewhat damning—to observe that the captain of the *Queen* had not given *City of Topeka*, nor had *City of Topeka* requested, compass bearings to the wreck. According to some sources, this was "an incredible display of poor seamanship" by the two captains.[9]

By now, both of the original shore parties, McCarthy's and Bunker's, were out of the picture. The McCarthy group was at the Cape Beale lighthouse, awaiting rescue, while the Bunker party was hunkered down in the lineman's cabin.

Acting in their place was a trio of men who had sought to come to *Valencia*'s aid—trapper Joe Martin, lighthouse keeper's son Phil Daykin, and lineman David Logan. They had hiked long miles and traversed a river in a borrowed canoe, and eventually they found a line atop the bluff and followed it; there, just below, was *Valencia*. For a moment, they—and the ragged, cheering survivors on the ship—thought that they'd be able to use the line to bring survivors ashore, but the line had snapped, worn and shredded after rubbing for hours on the sharp rocks. The parted line was useless.

Some sources, Neitzel included, take the three would-be rescuers to task, accusing them of not understanding—or not caring about—the urgency of the situation.[10] After all, the three testified that they had spent Tuesday night at the Klanawa River and then crossed over about 9:30 Wednesday morning. Later, according to their testimony, they stopped "for tea" at an Indian ranch. Why did they leave so late? Why stop for tea, of all things? The criticism may be unfair. In the first place, the three men had hiked for many hours, often cross-country, as they sometimes lacked a trail to follow during their roughly ten-mile trek. They were *exhausted*. They slept late—probably later than they intended. As for stopping for tea, some have noted that "tea," at that time and place, did not mean a leisurely sipping of a hot, restorative drink, as many Americans understand the term. "Tea," in Britain, Canada, and many other places, is a term used to mean an afternoon meal. One could

This image is typical of the rainforest underbrush found on Vancouver Island and the rest of the Pacific Rim National Park Reserve. This photo was taken in 2016; imagine the density of some areas of the island back in 1906, before many improvements had taken place.

argue pretty convincingly that the three tired men deserved a meal, and many have.[11]

Among those who felt that lifeboats *could* have made it to *Valencia* were the three men who stood atop the bluff overlooking the wreck site. Says one source, "All three men concurred in their opinion that lifeboats could have reached the steamer quite easily at this time."[12]

But no lifeboats were launched because by this time there were no ships left to launch them and no ships to collect people from those lifeboats if they *had* been launched, the *Queen*, *Salvor*, and *Czar* all having departed. *Topeka* was present, but she was steaming up and down the coast looking for survivors.[13]

The three men on the bluff looked down on a scene of desolation. There was no way down to the beach and no line they could use to bring survivors ashore. The cheering in the ship stopped, as the men and women marooned there suddenly understood that there was no hope for them. It would have been a stark and heartbreaking realization. The three would-be rescuers stared at the ship, only yards away, in dejected horror. Just after noon on Wednesday, January 24, they watched helplessly as the sea swallowed the SS *Valencia* and the last of her passengers and crew. Daykin later testified that as many as fifty people were left on board the ship when the ocean finally claimed it, most of them in the rigging.[14] No one left on board survived.

Chapter 22

The Ocean as Executioner

Not many of us have been thrust unexpectedly into a cold, dark, storm-tossed ocean, but given that most of us can swim, possibly quite well, we may find it a little surprising that a group of people can tumble into the ocean within sight of land and yet somehow not make it to shore. It's only yards away, after all. In a pool, you could easily swim that distance and, if you're in decent shape, not even get terribly out of breath. So it's hard to believe that so many of *Valencia*'s passengers and crew died so close to land. But it happened and, though it seems counterintuitive, the reality is that death was almost a foregone conclusion once they entered the water; there was almost no way for anyone to get from the foundering ship to a safe place on land, even though that land was so temptingly near. With a bit of background, we can see why it was almost suicidal for *Valencia*'s passengers and crew to attempt to swim to land once the ship had hit the rocks.

The temperatures on and near Vancouver Island, even in the winter, are actually some of Canada's mildest. For instance, the January temperatures far inland, say, at Toronto, Ontario, average about nineteen degrees F at night to a high of about thirty-one degrees during the day.[1] Meanwhile, over at Tofino, on the west coast of Vancouver Island, January temperatures range from a low of thirty-seven to about forty-six degrees; definitely chilly, but nowhere near as cold as on the mainland. In general, it's safe to say that January temperatures inland are almost always below freezing, while on Vancouver Island—except for in the mountainous regions—they are almost always above. The water just off the west coast of the island, then, tends to be well above freezing.

But water does not have to be freezing cold to kill. There are two dangers present when one is suddenly immersed in cold water. The first is known as *cold-water shock*. This occurs when one is suddenly exposed to water that is significantly below the body's normal core temperature. As one source notes, "Short of being hit by a bus or struck by lightning, cold shock is one of the biggest jolts that your body can experience."[2] Immersion in cold water almost always elicits a tremendous involuntary gasp from the victim; and that, not surprisingly, often results in an immediate drowning.[3]

If they survive the initial immersion, victims quickly tend to lose control of their breathing, resulting in hyperventilation and a feeling of suffocation. Hyperventilation can also result in a condition known as *hypocapnia*, a sudden reduction of the level of carbon dioxide in one's blood; this in turn can result in dizziness, faintness, cramping, and numbness—eventually leading to loss of consciousness.

In rough water, these issues are compounded, often by something that whitewater rafters have called *flush drowning*, which results when one has lost control of one's breathing and is at the same time repeatedly submerged by steep or breaking waves. It's simply not always possible to hold one's breath until it is safe to take a breath; the result is that one aspirates water instead of—or in combination with—air. In this situation, drowning is, once again, very likely.[4]

The other issue is *hypothermia*. It's fairly obvious that one cannot survive for long when one's body loses heat faster than the body can produce it. But what people often do not realize is that immersion in cold water is *especially* dangerous because water conducts heat away from the body twenty-five times faster than air.[5] So however quickly you might lose consciousness or control of cognitive and bodily functions due to cold on land, it would happen much more quickly in the water. (It's also important to note that children lose heat much more quickly than do adults; this is one more reason why no children survived the sinking of the *Valencia*.) While your normal body temperature is 98.6 degrees, hypothermia begins to set in once your body temperature falls below 95 degrees.[6]

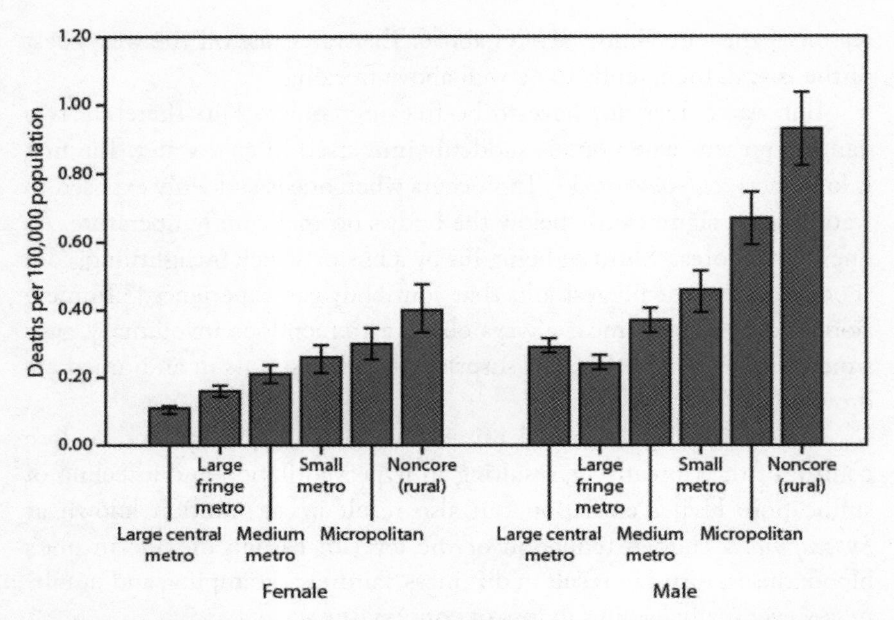

Death Rates Attributed to Excessive Cold or Hypothermia, by Urbanization Level and Sex. Note that deaths by cold or hypothermia are, understandably enough, much higher in rural areas and higher for males than for females.
COURTESY OF THE US DEPARTMENT OF HEALTH AND HUMAN SERVICES

So how cold is too cold? Some sources say that *cold shock* can occur when one is suddenly immersed in water anywhere between fifty degrees and sixty degrees. (It tends to vary based on how accustomed one is to sudden immersions in cold water.) While water of that temperature is considered extremely dangerous, even water temperatures of sixty degrees to seventy degrees are generally considered risky.[7] In other words, it's alarmingly easy to drown.

Off the west coast of Vancouver Island, say, at Long Beach and also at Bamfield, near where *Valencia* grounded, January water temperatures generally range from forty-five to fifty degrees.[8] This is well into the range of temperatures considered extremely dangerous. For children and for poor swimmers, young *and* old, a plunge into the waters off the coast of Vancouver Island in January was essentially a death sentence.[9] Even assuming that one is not battered against the rocks and has avoided aspirating water due to the tumultuous seas, a good swimmer—or a person using an effective flotation device—can be expected to survive for about three hours at most in water temperatures such as those encountered where (and when) *Valencia* went down.[10]

When we consider the frigid air temperatures, the cold water, the pounding seas, and the dangerous rocks, it's no wonder that few survived the wreck of the SS *Valencia*. The real miracle is that anyone lived at all.

Chapter 23

A Portion of Value

The arrival of *Salvor* and other vessels from Victoria, ostensibly to aid in rescue attempts, brings up a question that must have arisen in many minds as the ocean swiftly and steadily destroyed the *Valencia* and killed her crew and passengers: What was the likelihood that there were salvage opportunities associated with the stranded ship?

To examine that issue, we have to hopscotch through the history of salvage law, a complex and often obscure area of legal specialty.

Salvage law dates back to antiquity. Robert (Bob) Mester, of Northwest Maritime Consultants, notes, "In fact, salvage law is *so* old that it actually supersedes constitutional law."[1] So the basics of salvage were laid down long, long ago—and it's nowhere near as simple as many people believe.

"It's not a finders-keepers type of thing," says Mester. It turns out that just because you happen upon a derelict vessel, whether a small tender or a full-size ship, that does not mean you can just grab it and assume that it's yours by virtue of the fact that it's now in your possession. It's much more complicated than that. There are jurisdictional issues to iron out: What flag does the vessel fly? Where was it found? Perhaps most importantly, there's a certain amount—a great deal, actually—of due diligence that must take place to ensure that you actually have the right to claim the vessel or any materials that may have been on the vessel but are now floating free.

"Most such material, the vessel and its cargo, is or was insured," says Mester. That stands to reason because, looking back, there may well

Bob Mester and friend at work in the ocean
COURTESY OF ROBERT MESTER. USED WITH PERMISSION.

have been a time when a significant portion of a nation's wealth was transported on a fleet or ships, or even on one ship. So insuring a vessel and its cargo made sense, then and now. A finder cannot lay claim to materials simply because he or she found that material; a search must be made to ensure that the insurance company (or the original owner) has abandoned that claim. If so, you can lay claim to some portion, but not necessarily 100 percent, of the remaining value of the salvaged materials. (The cargo carried by the *Valencia* was said to be worth some $175,000—about $6 million today—at the time. It would certainly have been worth a salvage attempt, had any of the cargo remained intact.)[2]

And that really only deals with one piece of the law, the *law of find*. That is, it addresses what happens when someone finds a vessel or cargo seemingly abandoned. The other part of the law, the *law of salvage*, deals with the rescue of property that is in peril at sea. This is the set of admiralty laws that would have come into play during rescue operations—that is, while the ship was actually in danger. (Once that ship had broken up, it was no longer a ship in peril because it was in fact no longer a ship, just a collection of materials.) A *salvor* can reap a reward only if peril truly exists, and if the salvor can show that it played a role in recapturing the property that was on board. At that point, the salvors may find themselves awarded some portion of the value of the recovered cargo.[3]

That wording is important: "some portion of the value of the recovered cargo." *Valencia* carried cargo, most of it in the form of vegetables bound for markets in Seattle and in Alaska. By the time the sea entered the ship's holds—and certainly by the time the ship was destroyed—that cargo had little to no value; exposure to seawater would have destroyed most of the vegetables and reduced their value significantly. And the ship itself was mostly without value; broken up by the sea, there was little or nothing left to salvage.

At one point it was rumored that one of the two tugs, *Czar* or *Salvor*, had salvaged the ship's boilers as they lay in the shallow water among other detritus from the wreck. If true—the rumor was never substantiated—this would have been appalling for a couple of reasons. First, it would have come across as a sort of "looting" of what many had very quickly come to view as a sacred site, the place where so many had so recently lost their lives. Second, and perhaps more damning, if either or both of the tugs had gotten close enough to the wreck site to pick clean the bones of the ship, as it would no doubt have struck those following the story, it suggests that the vessels *could* have gotten closer to the wreck when they had responded in the first place. That raises questions about why they would not risk the shallow, rock-strewn water to save lives but were willing to do so in order to make a profit. That paints the two salvage vessels and their owners in a very unpleasant light, but perhaps unfairly, given that there was likely no storm raging during the time when such items might have been salvaged.

In any case, as Bob Mester points out, "It's unlikely that a vessel engaged in salvage would have bothered going out to a wreck site just to recover a couple of boilers; boilers are easy and fairly inexpensive to make—it doesn't seem to make sense that anyone would risk an operation to salvage an item that would be almost worthless and which could be quickly made in a nearby factory."[4] Professionals engaged in salvage are normally brave, fit, and intelligent; they're not prone to making foolish moves—that's why they're still alive and still in business.[5] Risking one's life to retrieve a couple of damaged boilers would have been a foolish move that would have involved a great deal of unnecessary risk for very little profit.

Of course, in British and Canadian common law, "salvors can also be rewarded for the salvage of *life*," says former Canadian Coast Guard Superintendent Clay Evans.[6] Thus, the vessels responding to the wreck could have benefited from having saved lives, as well as any cargo—not to mention the fact that the law of the sea, written and unwritten, requires one to attempt to render aid to vessels and people in distress, and sailors of all nations are known for doing so. We can assume that they would certainly have attempted to save the lives of the *Valencia*'s passengers and crew, if they had thought such an attempt was feasible.

In any case, while materials salvage may have been on the minds of some of the captains of the vessels that responded to the wreck, they would have quickly realized that there was going to be little left to salvage once the sea was finished with what had once been the SS *Valencia*. The true horror of the incident is that the only thing that *would* have been worth salvaging, the lives of the crew and passengers, is the one thing that was, in the end, abandoned.

PART III

The Aftermath

History is the story of events, with praise or blame.

Cotton Mather, *The Ecclesiastical History of New England*

Chapter 24

The Life and Death of the PCSSC

Companies are like people: possessed of the DNA of their parents (and in the case of companies, their investors), they live, they prosper for a time, and they die. Some of them live extraordinarily long lives. The Coca-Cola Company, founded in 1892 by Asa Griggs Candler and then sold in 1919 to the Trust Company of Georgia, lives still. Ditto IBM, founded in 1911 as the Computing-Tabulating-Recording Company and renamed International Business Machines in 1924; "Big Blue" remains an industry leader in emerging technologies and, as of 2022, holds somewhere around 150,000 patents.[1]

The Pacific Coast Steamship Company (the PCSSC, actually a subsidiary of the Pacific Coast Company), which owned the *Valencia* and many other vessels, did not have as prolonged a life as IBM, Nintendo, or Coca-Cola, but it was in business (under a few different names) for quite a while: It began in 1867 as Goodall, Nelson, and Perkins and became the Pacific Coast Steamship Company in 1876, after Christopher Nelson's retirement.[2] The company technically ceased to exist in 1916, when the Admiral Line bought its vessels, though the latter company continued using the nearly identical name, the Pacific Steamship Company; ironically, this is the name under which the company operated in the 1880s, long before its corporate dissolution. The Admiral Line ceased operations in 1936, so in one form or another, the PCSSC existed for roughly sixty-nine years.[3]

The Pacific Coast Steamship Company was a shipping giant: in 1906, the company owned some twenty vessels and did about two-thirds

A Pacific Coast Steamship flyer from about 1916. Note that two types of vessels are shown, a single-stack ship normally used for shorter voyages and a double-stack often used for longer voyages.
COURTESY OF HTTPS://WWW.TIMETABLEIMAGES.COM

of its business in freight, with the other third in passenger transport.[4] The number of passengers the line carried in 1906 is unknown, but in 1911, company vessels would transport almost 140,000 people.[5] But the company did not just run a steamship line—it also owned or invested in several railroads, was the shipping company of record for the Oregon Improvement Company (also owned by the parent company), and dabbled in publishing, timber, real estate, coal mining, and other concerns.[6]

Interestingly, in the semiofficial history of the company, *Ships and Narrow Gauge Rails: The Story of the Pacific Coast Company*, very little mention is made of the *Valencia*, although she is listed as one of the seventy or so steamships owned by the PCSSC or its parent company between 1867 and 1916. (*Valencia* was officially acquired by the company in 1897, though she had been on the West Coast for some years prior to that.) In the timeline presented in the book—which stretches from 1855, when the first wharf was erected at San Luis Obispo Bay, to the

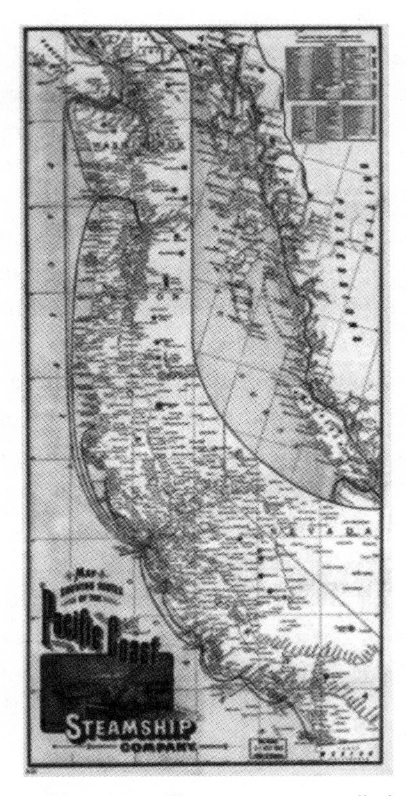

A map of PCSSC's steamship routes. The company excelled at marketing, printing shipping schedules, route maps, and myriad flyers—anything to get the company's name in the hands (and eyes and ears) of potential customers.

dissolution of the company's steamship interests in 1916—no mention at all is made of the SS *Valencia*.[7]

In sum, the Pacific Coast Steamship Company, in its various incarnations, was a large and powerful entity. It and its corporate parent, the Pacific Coast Company, employed a great many people, provided jobs for an even greater number not directly employed by the company, and was an influential organization in the Pacific Northwest, where it was headquartered during the *Valencia* years.

This is the reason so many doubted that local or regional investigators would be able to conduct an objective inquiry into the *Valencia*'s sinking and into the maritime community's response to that sinking. The PCSSC was not merely a company—it was a behemoth, a potent and sometimes domineering economic force that could wield a great deal of political influence at a time when local officials might well be under the sway of the men who held the reins at the company and who, one assumes, would prefer that the investigation be cursory, if there were to be an investigation at all.[8]

Thus, many agreed when newspapers in the Pacific Northwest (such as *The Morning Oregonian*, in its January 30, 1906, issue) worried in print that local investigators might be tempted to give the company a break for political reasons.[9] As a result, powerful businessmen (powerful businesswomen were few and far between at the time) and politicians from the Pacific Northwest and elsewhere around the country were moved to call for a federal investigation—which they got when President Theodore Roosevelt put his weight behind the creation of a commission of inquiry headed by then–Assistant Secretary of Commerce Lawrence O. Murray, as detailed in chapter 32. The federal commission, it was thought, was much more likely to conduct a thorough and objective investigation into the sinking of the *Valencia* and would be willing, it was hoped, to hold the Pacific Coast Steamship Company responsible for any misdeeds.

Chapter 25

Stacked Risks and Cascades of Failure

There are many questions associated with the *Valencia* disaster, and few satisfying answers. One of the biggest, most overarching questions is simply: *How could something like this happen?* The people involved were not stupid, not lazy, and not uncaring. They were professionals dedicated to doing their jobs as best they knew how and using the best tools available at the time. And yet seemingly everything that could go wrong ended up going wrong. Why? We tend to think of accidents as events that simply *occur*; we don't always attribute to them some causal agent. After all, they're accidents, right? They just happen. That's why we call them "accidents."

But that's a simplistic view. Every occurrence has a precedent, every effect a cause. It may be something obvious or it may be something obscure: if someone has too much to drink, gets into a vehicle, and runs into a storefront or another vehicle or even another person, we call that an "accident," but it's easy enough to point to the cause—somebody drank too much and attempted to operate a motor vehicle while impaired. Some "accidents" have causes that are less obvious, and some seemingly have *no* cause: When a strong wind blows one of your trees over and it falls onto your roof, that may look like what insurers call "an act of God," but there may be nothing godly about it. Perhaps you watered the tree too much, so its root system never gained a strong hold deep below the surface. It didn't have to; you watered it so often that it grew with a shallow root system and was therefore simply unable to stand up to all but the mildest of winds. In a case like this, *you* are ultimately the cause of your damaged roof, although that may not be apparent at first.

Risks cause "accidents." And when we multiply (or "stack") the risks, we increase the likelihood that an accident will occur. Simply put, risk stacking is a phenomenon in which the risk of an event occurring increases when multiple risk factors are present. It's a concept with which many systems experts are familiar.

Your family physician is in fact a "systems expert" because your body is a collection of systems—pulmonary, cardiac, neurological, and the like—that must function well in order for you to remain healthy. Your physician understands "risk stacking" quite well. That's one reason she looks at a variety of measures to estimate, say, your overall risk of a myocardial infarction—what we laymen would call a "heart attack." She makes note of your cholesterol level, your weight, and your participation in activities both healthful and unhealthful: Do you exercise, eat a low-fat diet, manage your stress levels? Or do you smoke? Eat lots of red meat? Lead a sedentary lifestyle? Are you in a high-pressure job? The physician will consider all of these and will, of course, examine you before estimating your risk of a cardiac "event." But whether you are a candidate for such an event or are likely never to experience one is largely up to you; though your genetic predisposition to various illnesses will certainly play a part, to a great extent, *you* have caused your heart attack—or the absence of one.

There are other forms of risk stacking, of course. *Financial risk stacking* can occur, for example, if you invest heavily in a single industry, have also taken on high levels of debt, and have a significant portion of your savings tied up in assets, such as real estate, that are difficult to liquidate. In this case, a combination of risks could lead to financial distress; there are many risks stacked together, so the likelihood of (often sudden) poor financial health has become more likely.

Risk stacking can be associated with another issue, *cascading failure*. This is a kind of failure in a system comprising interconnected parts, in which the failure of one part triggers the failure of successive parts. Such a failure is common in computer networks and power systems—and, not surprisingly, the human body. When multiple parts are interconnected and interdependent, the failure of one can create a "domino effect" in which downstream or related systems also fail. A power grid is a common and readily understandable example of this: If a station supplying power fails, the system is designed to balance the load among the remaining stations. However, this puts a strain on those downstream stations; because of the increased load, one may fail, causing others to fail, and so on.

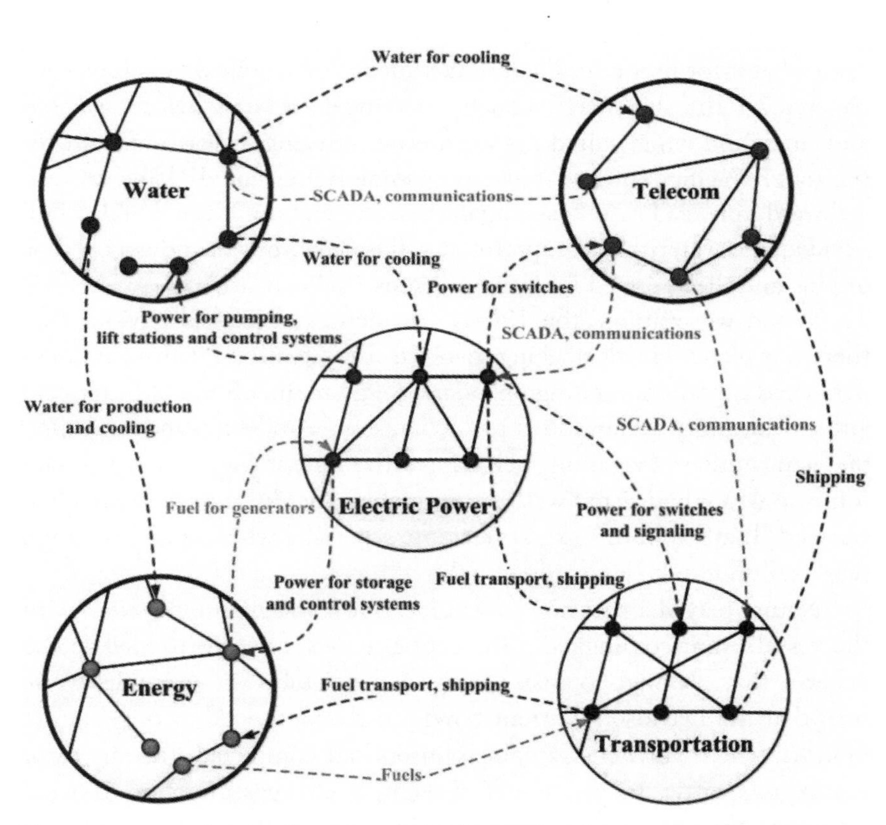

The main cause of cascading failure is the fact that systems contain interdependencies and are in fact often dependent upon other systems. This chart shows some of the interdependencies in various infrastructures, including water, electric power, transportation, and others. A failure in one could cause a disruption in all of them.

When we look at what went wrong with the *Valencia*, we can see the twin concepts of risk stacking and failure cascades in action. *Valencia* was a second-class ship under the command of a second-class captain. Oscar Johnson was by all accounts a good man, serious, knowledgeable, professional, and respected by his crew. But he was not an experienced shipmaster, had only made the San Francisco–Seattle run once before, and at a different time of year. He apparently understood neither the technology he used to navigate nor the currents in the area. He was somewhat error prone, and had in fact made costly mistakes before. (See chapter 7.) The ship itself, launched twenty-four years before, was old and tired, and while suited for warmwater cruising, it was not built for the tough Pacific Northwest waters. As one writer noted, "[She was an] awkward iron-clad vessel; lacking basic safety features like a doubled hull or adequate watertight compartments, the *Valencia* was under-powered, unsafe, and ill-equipped for the dangerous Pacific coast."[1]

When we examine the *Valencia* incident, it becomes obvious that there was plenty of risk stacking involved. The situation involved an inexperienced captain navigating on a dark, foggy night, on top of bitter cold and a wild, angry ocean, added to a reliance on—and a misunderstanding of—a primitive navigation technology, the patent log, which Captain Johnson wrongly insisted was overestimating the distance the vessel had traveled. That's a lot of risks to stack up, especially when sailing the high seas—an undertaking already fraught with risk.

Nature played a part here, of course. But it was nature *combined* with the vessel's shortcomings and the captain's inexperience that led to the debacle that claimed so many lives so needlessly. For example, we've seen that the Davidson Current flows mostly southward in the summer months, which is when Captain Johnson had completed his only prior voyage to Seattle. In the winter, though, it changes direction, flowing largely northward, thus pushing anything in its path—including the SS *Valencia*—directly at the rugged, rocky, and unpopulated west coast of Vancouver Island, British Columbia. (More about this current and Johnson's mismanagement of it in chapter 26.) This seasonal difference in the direction of the current was clearly noted on the charts of the day, but those charts—for reasons never adequately explained—were not carried on *Valencia*; the lack of charts contributed to the risk stacking and thus to the cascade of failure.

Because he failed to account for the current, and because Captain Johnson insisted that his log was overestimating (that is, "overrunning") the distance traveled by the ship, *Valencia* was about forty nautical miles farther north than he thought; by the time he was estimating that he would soon approach the easterly turn into the Strait of Juan de Fuca, he was long past the entrance to the strait and was bearing down on the rocks off the wild west coast of Vancouver Island.

To that volatile mix we add an inexperienced crew (or at least one lacking experience on that particular vessel), a paucity of navigational aids in the area, a lack of emergency and lifeboat drills, and an almost complete absence of nearby lifesaving infrastructure. Radio, which might have helped Captain Johnson determine his location—or which at least could have provided a means of calling for help—was not far enough along in its development to be carried by such small, commercial vessels.

Perhaps most importantly, the risk stack included a wild, inhospitable coast, one to which there was almost no safe access from the sea and from which there was no way to quickly get to any nearby settlements or, for that matter, to a stricken vessel; a ship lost on that rugged coast was *truly* lost, and there would be no help either from seaward or from ashore.

This was a stack of related, and in many cases preventable, risks that could and did erode any measure of safety that might have helped to ensure the survival of *Valencia*, her passengers, and her crew. The result was a cascade of failures, and those failures led to the dire results described in previous chapters.

Chapter 26

Deadly Reckoning

There's nothing sinister about ocean currents—or about the ocean, for that matter. It is what it is: cold, uncaring, and capricious. It simply carries on, and it does not care at all about the insignificant beings who attempt to transit or make a living from it.

Mind you, it's also not malign, though it can sometimes seem so; it's simply, as some have noted, "terrifyingly unpredictable."[1] It's not out to get you. But it is very powerful—perhaps the most powerful thing on Earth. It can kill you in mere moments, as it has killed many, but it does so not out of anger but merely because you happened to be in the wrong place at the wrong time and because you were not adequately prepared.

The captain and crew of the SS *Valencia* were not adequately prepared. They should have been; it was their business, after all, to cross oceans and to bring their cargo and passengers safely from one port to another. Ships and their captains have been doing so for thousands of years. But this captain made deadly mistakes, and foremost among them was his failure to understand the ocean currents in the area in which his ship was sailing.

There are, not surprisingly, people who specialize in oceanic currents and such phenomena.[2] They are called physical oceanographers, and one such person is Richard Thomson, a senior research scientist with Fisheries and Oceans Canada, the federal organization responsible for managing Canada's fisheries and ocean resources. Dr. Thomson points out that one of the main factors—though not the only one—in the creation of surface currents is wind. Having begun forming, currents in the

Northern Hemisphere are then forced by the Coriolis effect to the right of the direction of flow.[3]

"Near the coast of Vancouver Island, the Davidson Current merges with the Vancouver Island Coastal Current, which is a surface current that flows northward in winter along the west coast of North America from roughly northern California to southern Alaska," says Thomson. "The current is driven by the prevailing southerly winds associated with winter storms and establishment of the Aleutian low-pressure system in the Gulf of Alaska in winter."[4]

Off southwest British Columbia, the Davidson Current merges with the Vancouver Island Coastal Current, a year-round, northward-flowing nearshore current abetted by freshwater runoff that enters from the adjoining Salish Sea, says Dr. Thomson. People have known for many years about this particular current, although the mechanisms behind its formation have only been studied much more recently, beginning in the 1980s, by researchers such as Thomson and Dr. Parker MacCready and his colleagues at the University of Washington and elsewhere.

Thomson notes that the Davidson Current and the Vancouver Island Coastal Current "[are] more or less steady but can be reversed by very strong winds from the north, as occur in winter during times of high pressure and cold outflow winds from the continent."[5]

Thus, the Davidson Current is a known entity—and was familiar to mariners even back in 1906, when the *Valencia* embarked on what was to be her last voyage. In fact, it has been studied since the late nineteenth century, when it was named after George Davidson, a hydrologist working for the United States Coast Survey, the organization that eventually became NOAA, the National Oceanic and Atmospheric Administration.

In reality, the current is fairly active all year around, but—as is the case with many currents—it rises and falls: during the summer, it is found at great depths, around 650 feet; during the winter, though, it surfaces, affecting shipping to a much greater degree.[6] Because the now-surfaced current flows strongly northward in winter and is absent or much weaker, and may even reverse direction, in summer, it is important to take it into consideration when sailing in the waters of the Pacific Northwest.

Oceanographer Dr. Richard Thomson in his office
COURTESY OF RICHARD THOMSON

This is exactly what Captain Johnson failed to do. Or, viewed from another perspective, he did take it into account, but incorrectly, due to his inexperience. Having traveled this way before, during the summer months, Johnson assumed that the current was always weak and southerly—if it existed at all. He was wrong, and the reason he was wrong was that his previous journey from San Francisco to Seattle happened to have been his *only* journey from San Francisco to Seattle.

He then compounded his error by assuming—and by convincing some of the other officers on *Valencia*—that his patent log was overrunning, that is, overestimating the distance the vessel had traveled. The patent log is a simple but ingenious machine: a weighted rotor is dragged behind a ship. The rotor is connected to a clocklike mechanism that counts the number of revolutions the rotor makes in a given period of time. Given that number, one can calculate the speed of the ship based

A patent taffrail log used to measure the speed of a vessel and thus the distance the vessel has traveled. It works much like an automobile's odometer.

on the number of rotations and the time period. And given *that* number (that is, an estimated speed of the vessel), one can calculate the distance a ship has traveled: if the ship has traveled for one hour at 7 knots, for example, then the vessel has traveled 1.15 statute miles, or about 0.99 nautical miles. Essentially, a patent log is an aid used, and used to great effect, in dead reckoning.[7]

The problem is that the ship is not the only thing moving: the ocean moves too. As when aircraft pilots have to consider whether they're flying with or against a tailwind or headwind, a ship's navigator must take into account any ocean currents that may be working to speed up or slow down the vessel. In other words, the ship's speed as indicated by a patent log, because that simple machine cannot take into account any currents that may be hindering or helping the ship's advance through the water, may not register the ship's true speed over the bottom; the navigator must adjust his estimate of the ship's overall speed based on his knowledge of the currents in the area. This was not Captain Johnson's only error, but it is arguably the most critical mistake he made. Johnson disbelieved the patent log—and he was correct to do so. But he was correct in the wrong direction: the log was registering *too slow* of a speed, because the current was pushing his ship along much faster than he reckoned or than the patent log registered.

Because the ship was moving at a greater speed than he assumed, *Valencia* was much farther along her route than the captain anticipated; by the time he thought they should be approaching the area of the entrance to the Strait of Juan de Fuca, they had long passed that entrance—they were in fact much farther northwest than Johnson surmised, but the first he would know of this is when his ship slammed into the rocks off of Vancouver Island near Cape Beale, some twenty or more nautical miles northwest of the entrance to the strait. Oscar Johnson's misunderstanding of a straightforward and well-understood feature of the profession of which he claimed to be a master would cost an estimated 136 lives, his own included.

Chapter 27

Too Late, a Gift from Germany

Early twentieth-century mariners, like their predecessors, sailed the seas alone. Although their ships were occasionally in contact with others that passed near them (long-distance sailors would sometimes meet up with passing vessels for "gams" when they "spoke"—that is, spied and made contact with—another ship), most sailors, once at sea, were in contact with no one who was not aboard their vessel.[1] It could be a lonely life, and sailors sometimes went for months or even years with little or no word from home. Hiding ships, that is, vessels sent from New England to the west coast of the United States and Mexico to collect and bring back hides and pelts, could remain away from home for two, three, or even four years. Whalers could remain on the hunt for similar periods of time.[2]

As we've seen, radio—if it had existed more widely at the time—would have made *Valencia*'s brief voyage both safer and less lonely. The voyage must have struck some on board, those who knew something about these marvelous new Marconi radio sets, as even more solitary than necessary, given their realization that *some* ships were in fact in more or less constant contact with land stations and with other ships; some were even allowing passengers to contact their businesses and their families while at sea, so it must have upset at least the more informed among them not to have access to this amazing new technology, even for what was meant to be a short voyage.

Perhaps more importantly, radio might have made the *Valencia*'s run to Seattle *safer* because the officers on board would have been able to

The radio room (or "shack") aboard the SS *Titanic* during its maiden (and only) voyage. Radioman Harold Bride, who survived the sinking, is seated before the apparatus.

confer with others about their location. And when disaster *did* strike, radio might have allowed them to summon help more quickly, although the presence of radio might not have helped very much in this case, due to the circumstances of their stranding.

It's important to realize that those in positions of power—that is, those who held the governmental purse strings that might have helped address the many losses associated with wrecks on or near the coast of Vancouver Island—were not unaware of advances in radio and how those advances might have been put to use; they simply didn't care. In 1905, for example, three months *before* the *Valencia* tragedy, the writer of a letter to the *Colonist* finds it lamentable that the notion of "wireless telegraphy has been left practically out of our list of [safety-related] propositions," commenting on the large number of "vessels on which the wireless apparatus would have been extremely useful."[3]

In 1906, there were other developments on the horizon in the embryonic world of electronics that also could have helped save *Valencia.* One of these actually made its first appearance two years *before* Captain Johnson and his crew embarked on what would turn out to be the ship's last voyage.

In Dusseldorf, Germany, in the early 1900s, engineer Christian Hülsmeyer was experimenting with reflectors inspired by Heinrich Hertz. (Hertz, a nineteenth-century German physicist, had himself conducted experiments aiming to verify the still-earlier work of Scottish physicist James Clerk Maxwell, who had predicted that light and radio waves were in fact similar, but of different frequencies.) Hülsmeyer discovered that electromagnetic waves emitted by a transmitter and then reflected by a metallic surface could be used to detect distant objects.

Eigentum des
Kaiserlichen Patentamts.
Eingefügt der Sammlung
für Unterklasse ....
Gruppe Nr. ....

KAISERLICHES PATENTAMT.

AUSGEGEBEN DEN 21. NOVEMBER 1905.

PATENTSCHRIFT

— № 165546 —

KLASSE 74d.

CHR. HÜLSMEYER in DÜSSELDORF.

Verfahren, um entfernte metallische Gegenstände mittels elektrischer Wellen einem Beobachter zu melden.

Patentiert im Deutschen Reiche vom 30. April 1904 ab.

Vorliegende Erfindung hat eine Vorrichtung zum Gegenstande, durch welche die Annäherung bezw. Bewegung entfernter metallischer Gegenstände (Schiffe, Züge o. dgl.)

Der Apparat besteht, wie bei der drahtlosen Telegraphie, aus Gebe- und Empfangsstation, nur mit dem Unterschiede, daß beide sich an demselben Punkte befinden, allerdings

Christian Hülsmeyer's 1904 German patent for a metal detector that would eventually become RADAR

He then took what would turn out to be a very significant step: he found a way to measure the *time* it took to reflect those waves. And once that had been accomplished, it was fairly simple—knowing the speed at which light travels—to determine the *distance* between the transmitter and the reflecting object. On April 30, 1904, Hülsmeyer patented his discovery in his native Germany.

The next step in this process was a series of experiments intended to verify the effects predicted by Christian Doppler back in 1842. Doppler, an Austrian physicist, had theorized that radio waves would change in frequency as an object reflecting them moved. When you hear an ambulance drive by with its siren blaring, the peculiar change in pitch that occurs as it passes is due to the Doppler effect. All kinds of wave phenomena—sound waves, light waves, radio waves—shift in frequency when the source of the wave approaches or recedes from the point of observation. When you view a "Doppler Weather Forecast," that forecast has made use of the changing reflected frequencies in order to note the distance, speed, and direction of a storm front or other meteorological phenomenon.

Once this had been accomplished, one could improve on Hülsmeyer's device in order to determine whether the object reflecting those signals was stationary or moving, and if moving, the direction and speed of its travel.

One could now *range* a distant object (calculate its distance from a known point) and determine its direction. Hülsmeyer, with some help from Hertz and Doppler, had just invented RADAR: radio detection and ranging. Hülsmeyer called it, rather clumsily, a "telemobiloscope," and he actually built one and showed it to the German navy, which professed profound disinterest at the time. (They would change their tune when World War II rolled around.) Nonetheless, only six days before *Valencia* left San Francisco, Hülsmeyer's invention was granted US patent number 810,150.[4] RADAR was about to become a reality, though it would be greatly improved in the 1930s and 1940s, due to advances fueled by World War II, but—as with shipboard radio and newly built lighthouses and lifesaving stations—too late to do *Valencia* any good. (Note that the acronym has by now become so familiar that few bother to capitalize it and even fewer recall that it actually *stands* for something; it's so widely used that it has become—or is in the process of becoming—a word, as sometimes happens. Consider terms such as "SCUBA" and "AIDS," and also words such as "motel" and "hazmat," the latter two of which are more properly referred to as blends. Not many recall that a motel is really a "motor hotel," or that SCUBA refers to a self-contained underwater breathing apparatus, or that hazmat stands for "hazardous materials.")

Think what might have happened had *Valencia*'s crew been able to use radar to peer through the dark and the rain and the fog to "see" the Vancouver Island coastline looming ahead in the inky darkness; or, more to the point, think what might *not* have happened. Perhaps the captain would have been given warning early enough to have backed off and sailed back out to sea until he could figure out his location, possibly waiting for daylight and the storm's passage to confirm his safest course of action. Of course, this is what he should have been smart enough to do even *without* radar, but perhaps with radar he would have taken heed much earlier.

It's unfortunate, of course, that *Valencia* had neither radar, nor radio, nor any other technological advantages that might have resulted in a different outcome. Then again, had she a different master, the entire debacle—not all of it his making, admittedly—might have been avoided in the first place.[5]

Chapter 28

New Lighthouses and Lifesaving Stations

It's an unfortunate fact of life that high-profile disasters are often necessary catalysts for much-needed change; apparently, there are times when people have to die so that others may live. The sinking of the SS *Valencia*—and the largely unnecessary deaths of roughly 136 passengers and crew—did at least some good, spurring the US and Canadian governments to address shortcomings in maritime safety policies and infrastructure in Pacific waters.

Of course, citizens from both countries had been calling for more and better safety measures along the coast for many years. As maritime historian and former Canadian Coast Guard Superintendent Clay Evans notes, "Although the death toll from the *Valencia* incident was large . . . this was by no means the worst maritime disaster along the coast of British Columbia up to this point. Calls from both sides of the border for more maritime safety measures had been loud and clear for several decades and the carnage of ships, cargoes, and souls was never ending."[1]

Eventually, both governments heeded those calls, to varying degrees. Perhaps the most obvious example was Canada's 1907 construction of the Pachena Point lighthouse on the west coast of Vancouver Island in British Columbia. The lighthouse, the presence of which might have alerted *Valencia*'s officers to the dangerous rocks they were approaching, stands about an hour or so south-by-southwest of Port Alberni, British Columbia, and about thirteen miles north-by-northwest of the First Nations village of Clo-oose. The lighthouse's website notes that "the seas surrounding Pachena Point are littered with the wreckage of scores of ships" and comments that the sinking of *Valencia* was one of the primary reasons for the construction of the lighthouse, but does not attempt to explain why it took so long for the government to decide to build one there.[2]

Located at one end of what is now the West Coast Trail in Canada's Pacific Rim National Park Reserve, the Pachena Point lighthouse stands on an octagonal wooden tower some 153 feet above the mean high-water mark and can be seen for fourteen nautical miles. These days, it flashes every 7.5 seconds. These are the bare facts about the lighthouse, but what we don't see, unless we're willing to look very carefully, is the pain, terror, and death—the terrible *loss*—that lies behind the origin of the lighthouse. Between the years 1800 and 1907, hundreds of people died in shipwrecks near the point where the lighthouse would someday be located. The *Valencia*'s sinking was the final straw: bowing to public opinion, the Canadian government finally determined that a lighthouse near Pachena Point was a necessity; of course, Canadian and American citizens had been telling them that for many years.

A few years later, in 1912, the Canadian government built the Sheringham lighthouse, in Shirley, British Columbia, farther up the strait and about an hour or so west of Victoria. The Sheringham lighthouse is directly across from Twin Beach, Washington, on the American side of the strait.[3]

In the years after the wreck of the *Valencia*, Canada's Department of Marine and Fisheries, now known as Fisheries and Oceans Canada, was intent on improving its system of aids to navigation all along the Strait of Juan de Fuca, largely in order to safeguard the shipping on which so many industries in the area relied, including logging, mining, and canning concerns.

Today, there are eight lighthouses dotting the western coast of Vancouver Island, ranging from the Nootka and Cape Scott lighthouses in the far north and west to the more southeasterly ones at Sheringham Point and the Race Rocks lighthouse. In 1906, both the Carmanah and Cape Beale lighthouses existed, but neither of them would have been any help to *Valencia*, as she crashed into the rocks near Pachena Point and never sighted them.

The lighthouse at Pachena Point came online in 1907, largely as a result of the *Valencia* tragedy.

Even with the additional lighthouses, British Columbia has fewer than any other Canadian coastal province, except for Manitoba, which has only two. (Then again, Manitoba has a much smaller coastline to protect than the other provinces, and thus correspondingly less danger and fewer shipwrecks.)

The Sheringham Point light was erected in 1912, another response to the sinking of the *Valencia* and other vessels off the coast of Vancouver Island.

Two lighthouses, at Cape Beale and at Carmanah Point, were active on Vancouver Island when *Valencia* struck the rocks, but the first was too far north and the second too far south to be of any use.

In addition to more and better-located lighthouses, the wreck of the *Valencia* served to call attention to the need for various forms of lifesaving stations in the area, and the government of Canada addressed the issue by building and staffing stations at Pachena Bay itself; at Tofino, northwest of Pachena Point; and at Victoria, far up the strait from Pachena Point and directly across from Port Angeles, Washington, on the US side.

In addition, a station was established in 1907 at Bamfield, roughly eight miles northwest of Pachena Point.[4] The new Bamfield station included a boathouse and a track for launching and retrieving the station's (at first oar-powered) lifeboat.[5] Perhaps more importantly, in terms of keeping a crew at the ready, it was decided that crew stationed at Bamfield were to be paid $40 per month, plus a mess allowance of 60 cents per day.[6] In an era in which many Canadian stations were staffed by unpaid volunteers, this was a much-needed move toward professionalizing the stations and paving the way for the establishment of what would eventually become the Canadian Coast Guard.

Another early response to the *Valencia* tragedy was the establishment of a lifesaving station at Pachena Bay. Prior to the establishment of the Pachena Bay station, Lloyd's of London and the Canadian government had agreed to keep the steamer *Salvor* ready to send out in the event that a rescue became necessary. However, the steamer was stationed at Esquimalt, some distance away from the rougher, more volatile waters on the coast of Vancouver Island, so it was determined that, while helpful, that was not enough: permanent stations needed to be established closer to where accidents were likely to occur.[7] And as we have seen, the presence of *Salvor* and *Czar* did little to alleviate the suffering of those on the doomed *Valencia*, in any case. Similarly, in 1908 a lifesaving station was established at the northwestern entrance to Barkley Sound, near the village of Ucluelet.

Efforts to improve maritime safety on the west coast of Vancouver Island continue today, with a new search and rescue center built in 2020 at Tahsis, in the northern portion of the island. The station houses "a 15-metre Coast Guard lifeboat and a rigid-hulled inflatable boat, providing 24-hour availability for search and rescue and environmental response missions." Three other new stations were planned for Victoria, Hartley Bay, and Port Renfrew.[8]

On the US side, less has been done. In 1878, the US Congress had passed the Life-Saving Service Act, which created the US Life-Saving Service. After the *Valencia* incident, the government undertook to improve the organization, allocating additional funding and taking steps to further professionalize it. (Eventually, the US Life-Saving Service merged with the Revenue Cutter Service to form our modern US Coast Guard.) The US Life-Saving Service built a network of stations, most along the East Coast, and equipped them with surfboats, breeches buoys, and other rescue equipment, but there were few in the Pacific Northwest.

The Cape Flattery light was there all along (or at least since it was first activated in 1857), though in the swirling fog and stormy weather it was missed by *Valencia*'s officers and crew. Since 1906, only two more lighthouses have been built on the US side along the strait in the area commonly known as the Olympic Peninsula. Another US lighthouse, in operation since 1857, is located five miles out on the Dungeness Spit, north of Sequim, Washington.[9] One other lighthouse, located near Port Townsend, Washington, was built several years after the wreck of the *Valencia*.

Additionally, a lifesaving station near Neah Bay, not far from the Cape Flattery lighthouse, was soon authorized by Congress, which allocated some $30,000 for equipment (over $1 million in today's currency), including two self-righting, self-bailing lifeboats. In addition, the authorization also required "that there shall be constructed, for and under the supervision of the Revenue-Cutter Service, a first-class ocean-going tug, for saving life and property in the vicinity of the north Pacific coast of the United States, which said tug shall be equipped with wireless-telegraph apparatus, surfboats, and such other modern life and property saving appliances as may be deemed useful in assisting vessels and rescuing persons and property from the perils of the sea at a cost not to exceed one hundred and seventy thousand dollars."[10] (The Revenue Cutter Service, established way back in 1790, was a precursor to the US Coast Guard;

it operated for 125 years, prior to the official establishment of the latter organization in 1915.)[11]

These days, of course, with ubiquitous radio communications facilitating rescue efforts in the Pacific Northwest and almost everywhere else, the various lifesaving apparatuses of both countries are in constant touch, so efforts can be coordinated and rescue attempts mounted almost

The *Santo Cristo de Burgos* was a Spanish galleon built in 1687–1688. It sank with its cargo of beeswax around 1693 near the northern coast of Oregon. This ship, based at the Maritime Museum of San Diego, is a replica of a typical galleon.
IMAGE PLACED IN THE PUBLIC DOMAIN AND USED UNDER THE CREATIVE COMMONS CC0 1.0 UNIVERSAL PUBLIC DOMAIN DEDICATION

instantaneously. That, added to the other lifesaving infrastructure put into place since the sinking of the *Valencia*, has helped greatly reduce the number of shipwrecks in and around the so-called graveyard of the Pacific.

However, even those improvements in maritime safety infrastructure have not completely eliminated the danger in and around British Columbia. Several ships have sunk off the coast there during the years after 1914—an admittedly arbitrary point in time, but one by which additional lighthouses and lifesaving stations had been established and radio communications had improved. These include the *Clarksdale Victory* in 1947, the US Army tug *Major Richard M. Strong* in 1949, the tug *Alberni* in 1915, the tug *Rogue* in 1975, the *New Carissa* in 1999, and the SS *Iowa* in 1936.[12] (Possibly the *first* ship to be lost in the area was the *Santo Cristo de Burgos*, which sank in or around 1693 off the Oregon coast. The ship was carrying a cargo of beeswax—very valuable at the time—and clumps of beeswax would occasionally show up on beaches and in the hands of native traders for hundreds of years. In fact, early settlers in coastal Oregon mined the buried beeswax blocks and sold them. Some of the beeswax is still exhibited in museums in Tillamook, Oregon, and elsewhere.)[13]

Chapter 29

An Illusory Path to Salvation

Vancouver Island's West Coast Trail is brutal. If you're planning to hike its seventy-five-kilometer (forty-seven-mile) length (a trek that, from end to end, normally takes six to eight days), Parks Canada recommends that you bring along the following gear: sturdy boots, rainwear and warm clothing, a backpacking stove and fuel, a backpack (preferably one lined with plastic bags) that features a padded hip belt, a synthetic-filled sleeping bag (down bags lose their warmth when wet), a sleeping pad, a trail map, tide tables, fifty feet of synthetic rope, an emergency signaling device, toilet paper, water purification equipment, waterproof matches, a flashlight, a weather radio, and lightweight shoes for river crossings and for wear in camp. (The Parks Canada website also warns you *not* to bring firearms, an axe, or any pets.[1]) You can tell by the recommended clothes and gear that weather is a concern on the trail, as are river crossings, mud, rockfalls, downed trees, and other potentially dangerous obstacles. In addition, the trail features roughly seventy ladders, some as tall as thirty feet, that must be scaled in order to move to the next section of trail. As the WCT website notes, "It is sometimes funny and sometimes worrying to see hikers coming toward you with a look of agony and determination. A very fit person will find the trail difficult."[2] The trail is open only by reservation and only from May 1 to September 30. It is rated one of the world's top hiking trails and is demanding enough that roughly 75 of the 7,500 people who attempt the hike every year end up having to be saved by search-and-rescue professionals.[3]

Before Parks Canada took it over in 1973, well before Parks Canada existed, in fact, the trail was known as the Dominion Lifesaving Trail.[4] It had been built (or expanded, at any rate) in 1907 to help provide rescue for anyone shipwrecked along the turbulent coast.[5] The prime impetus for the expansion of the trail was the sinking of the *Valencia*, although locals had lobbied for years for trail improvements meant to help shipwrecked sailors and others lost in the rugged wilderness of Vancouver Island's west coast; anyone stranded on the rocky coast, went the reasoning, had nowhere to go but up to the top of the bluffs, and they needed a half-decent trail there if they were going to hike out to civilization.

Before *that*, the trail was simply a barely visible path that followed the telephone and telegraph wires strung in trees along its length. Before that—*long* before that—the island's Native peoples had used the trail,

A view of the Pacific coast from the West Coast Trail. Note the absence of any actual beaches; this is typical of many spots along the Vancouver Island coastline and the Pacific Northwest in general.

USED UNDER THE CREATIVE COMMONS ATTRIBUTION-SHARE ALIKE 3.0, COURTESY OF WIKIPEDIA USER JONATHAN.S.KT.

and others like it, for some four thousand years before the Europeans came.[6] In the early 1900s, one might encounter a simple lineman's hut every few miles, placed there to provide shelter for those who maintained the miles of wire that kept the far-flung outposts of Vancouver Island's governmental infrastructure (Cape Beale and Victoria, for example) in touch during emergencies. Shipwrecked sailors might occasionally find shelter in those huts—if they could locate the huts.

These days, the trail is part of the Pacific Rim National Park Reserve, a preserve comprising almost two hundred square miles of near wilderness located entirely on Vancouver Island and other nearby islands. (Interestingly, the boundaries of the park actually extend quite a distance offshore, in order to protect the marine environment associated with the area.)[7] As part of the agreement that created the park, there are no forestry rights allotted within the park, so logging and other such industries are prohibited. Nonetheless, the trail has been widened a bit, and much of the underbrush along the trail itself has been cleared since 1906, when the *Valencia* struck the rocks just off the desolate coast, so the going is not quite as tough as it was when the *Valencia* shore parties were struggling to make their way through the forest.

It is this trail, or its precursor, that the Bunker and McCarthy parties used to find their way to civilization. (Or, at least, to points along the trail where they could use the telephones in the linemen's huts to *contact* civilization.) As was pointed out elsewhere, if the two parties had actually stayed near the ship, someone would almost certainly have been around to capture one of the lifesaving lines fired from *Valencia*, which could then have been used to bring the remaining passengers and crew ashore. One of the members of the McCarthy party, John Marks, almost certainly knew better than to leave the site. Marks was a seaman on the ship, an experienced sailor; he and McCarthy—an even *more* experienced seaman—both should have realized that the best way to save the people left on board was by retrieving a line fired by the Lyle gun. Instead they went off into the forest, looking for help. In their defense, the overgrown forest made passage back to the ship difficult, and the Cape Beale lighthouse was only a few miles away.

However, both parties left the scene, the Bunker party heading west and the McCarthy party heading east, the latter group having landed "about seven miles northwest of the wreck" and apparently unable to work their way through the dense forest back to the ship.[8] Both followed the trail and both managed to find lineman's huts, where they used telephones to call for help and to get the word out about the sinking. Many researchers, as well as the official investigators, are of the opinion that the two parties' leaving the area and striking out on the trail doomed the *Valencia* survivors, even though their intent—finding help—might have been laudatory.[9]

Even the official investigating commission, which took great care to avoid levying blame disproportionately, hastily, or unfairly, notes in

Parks Canada often leaves a set of red Adirondack chairs at important, meaningful, or especially beautiful spots in their parks. These chairs are at Silverhorn Creek at Banff National Park in Alberta, Canada. Two similar chairs are placed at kilometer 18 along the Pacific Coast Trail; those mark the destruction of the SS *Valencia*, which took place just below the trail at that spot.
PHOTO BY LESLEY SCHER. USED WITH PERMISSION.

its final report that the two shore parties—especially the Bunker party, which was only about half a mile from the wreck when the members reached the top of the bluff—were the ship's best, and perhaps only, hope.[10] When the Bunker party turned west and left the *Valencia* to its fate, says the official report, "it must be said that by far the best chance for rescuing the remaining survivors on the *Valencia* vanished."[11] The report goes on to note that the distance between the stern of the *Valencia*—recall that Captain Johnson had pivoted the vessel such that her stern now pointed toward the shore—and the top of the cliff was probably not over 250 feet, easily within reach of a line fired by the ship's Lyle gun. Lines were in fact fired by the increasingly desperate crew, but there was no one on shore to pick up a line and make it fast; any lines that did make it up to the top of the cliffs simply lay there until they parted under the strain and the chafing of the rocks and shrubs. Several times, a literal lifeline waited in the dirt, unnoticed and unused; dozens of people died because there was no one at the top of the bluff to grasp a piece of rope.

Chapter 30

The First Investigations

Shortly after the *Valencia* disaster—in early February 1906—President Theodore Roosevelt ordered the Assistant Secretary of Labor Lawrence O. Murray (more about him shortly) to convene a federal investigation into the wreck. This federal inquiry is the best-known investigation of the wreck of the *Valencia*, and it is widely quoted, but it is not the only inquiry; it's not even the only US investigation, nor is it the earliest one.

In fact, there were multiple investigations, the first one commencing on Saturday, January 27, 1906, only four days after the ship was lost, at a time when bodies were still washing ashore. That Seattle-based inquiry was conducted by Captains Bion B. Whitney and Robert A. Turner, of the Steamboat Inspection Service. The two men, the same two who had helmed an inquiry into the disastrous fire on *Queen* two years earlier, relied on testimony from the thirty-seven survivors to help describe the accident and its causes.[1]

As might be expected, though, that investigation was criticized by those in the community who felt that a local—or even regional—commission could not truly be objective and would not be able to guarantee a comprehensive and impartial job of looking into the matter. (Especially given the political and economic clout wielded in the area by the Pacific Coast Steamship Company.) Thus, even as Whitney's and Turner's investigation proceeded, influential Seattle citizens were in touch with representatives of the federal government, imploring them to undertake a more wide-ranging—and, one assumes, more objective—inquiry. (That federal investigation will be discussed in the next chapter.)

The main issue with the Seattle-based investigation was that a local inquiry was presumed to be tainted by political influence. The *Seattle Times*, not then known as a beacon of impartiality (or journalistic decorum, for that matter), trumpeted in a subhead, "Valencia Investigation Practically Conducted by the Pacific Coast Steamship Company and Its Employees."[2]

Nonetheless, Secretary of the Department of Commerce and Labor Victor H. Metcalf made it known to the local inspectors that he expected from them a "thorough and searching investigation of *Valencia*; also full investigation of conduct of officers of steamers *Topeka* and *Queen*,

The Steamboat Inspection Service was not officially created until 1871, but safety inspections of US-flagged merchant vessels have been mandated since 1838, as noted on this relief, sculpted on the facade of the US Department of Commerce building, Washington, DC.

pursuant to Section 4450 Revised Statutes." He also ordered the inspectors to "investigate all causes of wreck, the loss of life and any misconduct or neglect of duty on the part of any connected therewith." Metcalf also insisted on public hearings and a "thorough and complete" report.[3]

That early inquiry concluded on February 13 with the investigators' final report to the Department of Commerce and Labor dated March 17, 1906. The investigation is said to have generated over one thousand

Secretary of the Department of Commerce and Labor Victor H. Metcalf, shown here in 1905, would be named Secretary of the Navy shortly after the *Valencia* incident.

pages of transcribed testimony but "added little to an understanding of the tragedy."[4]

The "thorough and complete" report requested by Secretary Metcalf was indeed delivered in a timely fashion, but its findings, labeled "Preliminary," in any case, were almost immediately superseded by the federal investigation's report and by a Canadian investigation. (Though the conclusions of the latter seem to have mysteriously disappeared. More about that momentarily.)

Valencia, though an American vessel carrying mostly American passengers to a US destination, went down in Canadian waters, and the rescue operations, abortive as they turned out to be, were to some extent Canadian in origin. The Canadian government therefore established a Commission of Enquiry on February 6, 1906. Chairing the investigating committee was Captain James Gaudin, of whom we have previously read. Gaudin, a seafaring man of great experience—and the past master of several ships—had played a major part in many previous marine investigations. He was a local agent of what was then known as the Canadian Marine and Fisheries department, a well-respected officer, and in spite of being a bureaucrat, was credited after the investigation with having played a key role in ensuring that the Canadian government acted upon the committee's recommendations to establish more, and better, maritime safety aids on and off the coast of Vancouver Island, British Columbia.

Nonetheless, the commission's report is nowhere to be found. As maritime historian Clay Evans put it, "once the draft recommendations were completed and forwarded to . . . the minister in Ottawa, it was never seen again."[5]

Evans believes that Captain Gaudin, who apparently had an awkward habit of "speaking truth to power," asserted that the Canadians—both in the form of the ships and personnel that had been dispatched to the scene of the *Valencia* wreck and in terms of the woeful absence of navigational aids and lifesaving equipment in the area—bore much of the responsibility for the disaster. Evans feels that the commission's, and thus Gaudin's, conclusions "were not repeatable in the bastions of authority in far-away central Canada."[6] The discomfort with which some of the

VETERAN COAST NAVIGATOR DEAD

Captain James Gaudin Passes Away at His Victoria Home.

Was Wreck Commissioner for British Columbia, and Well Known.

Captain Gaudin would pass away in Victoria, British Columbia, on January 13, 1913, at the age of seventy-five, almost exactly seventeen years after the *Valencia* sinking.
FROM *THE DAILY PROVINCE*, OF VANCOUVER, BC

report's conclusions may have been viewed in the corridors of power may have led to its disappearance.[7]

"I think that the [Canadian] report was purposely 'disappeared,'" says Clay Evans. "I'm pretty sure that Captain Gaudin, having 'spoken truth to power,' and being a strong advocate of more resources . . . I think it probably died in Ottawa. I think they heard it, but they didn't want it to go up [the chain of command] any more." Evans notes that the Canadian government was actually pretty fastidious about keeping records, and he was therefore surprised, having visited the National Archives, that there is absolutely no evidence of such a report in those archives.[8]

Nonetheless, even in the absence of the report itself, we do have, thanks to a 1907 issue of the *Vancouver Province*, a recap of many of the report's conclusions. Some of those fly in the face of both common sense and the official testimony. For instance, the report apparently stated that the Canadian commission was unable to ascertain whether the then-present aids to navigation "were sufficient to prevent a recurrence" of similar incidents in the future, partly because none of the *Valencia* officers had survived to provide firsthand accounts of the tragedy. As Evans points out, that is simply not true—the *Valencia*'s second officer, Pete Peterson, not only survived but testified to the US investigators.[9] In addition, it was obvious on its face that there were not enough aids present: there were no nearby lifesaving stations, no lighthouses near the wreck site on the west coast of Vancouver Island, and no rescue vessels stationed near enough to get to the site within a reasonable amount of time. And as Captain J. W. Troup, one of the Canadians aboard *Salvor*, noted somewhat sardonically, "a self-righting, self-bailing life-boat might have done some good that day." Meanwhile, Pacific Coast Steamship Company Captain J. B. Patterson castigated the Canadians for not having put in place "some thoroughly good life-boats . . . and a properly equipped station and a crew of paid men, not volunteers."[10]

Nonetheless, several of the Canadian report's recommendations made sense and were in fact adopted by the commission, including the construction of a radio-equipped lighthouse at Pachena Point, an extension of what had come to be called a "life-saving road" (now part of the West Coast Trail described in chapter 29), dedicated lifeboats stationed at several points along the coast, and the building of new stations at Bamfield and Ucluelet.

Of course, the mere adoption of such measures was not, in and of itself, especially useful. The measures needed to be *funded* in order to have any real effect, and according to Evans, "the Department of Marine and Fisheries continued to drag its heels in providing funding for its own plans and recommendations. This despite the inquiries, the decades of lobbying from all corners, and even the constant solicitation from their own marine agent, Captain Gaudin, for effective action."[11]

A young James (J. W.) Troup, long before becoming a prominent and respected steamboat captain
FROM *LEWIS & DRYDEN'S MARINE HISTORY* OF THE *PACIFIC NORTHWEST*

Many of the recommendations were eventually acted upon, but it took years of patient (and occasionally impatient) lobbying from Gaudin and others to see to it that the government of Canada lived up to its promises. The country's lack of preparation was a national disgrace, and its continued equivocation when it came to actually enacting the proposed maritime safety changes in British Columbia was a further embarrassment.

Chapter 31

"Cold-Blooded Murder"

By far the most thorough—and most frequently quoted—official investigation was the US one, which came about when, on February 6, 1906, Congressman William Ewart Humphrey of Washington called on President Theodore Roosevelt to appoint a special commission to investigate the wreck.[1] Humphrey believed that the Seattle inspectors were not competent (or perhaps not objective enough) to handle the investigation and that the Pacific Coast Steamship Company exerted too much influence over the inquiry. The following day Roosevelt ordered Secretary of Commerce and Labor Victor H. Metcalf to create a "Federal Commission of Investigation" to hold independent hearings and to make a complete examination of all the circumstances surrounding the wreck of the *Valencia.*[2]

Metcalf then ordered Lawrence O. Murray, assistant secretary of Commerce and Labor, to head the investigation. Murray was to focus not only on the causes of the wreck but also on prevention and navigational safety along the coast and inland waters of Washington State, as a means of preventing future such tragedies. Murray was appointed chairman of the commission, and Herbert Knox Smith, deputy commissioner of corporations, and Captain William T. Burwell, US Navy, commandant of the Puget Sound Navy Yard, were also members. The commission commenced the investigation in Seattle on February 14 and concluded on March 1, 1906. The commission examined sixty witnesses and collected 1,860 pages of testimony and more than thirty exhibits.[3]

The special commission meeting in 1906. Unfortunately, it's difficult to tell who's who in this unlabeled, uncaptioned photo.

In addition to hearing testimony, the commission members boarded the lighthouse tender SS *Columbine*, which took them to Neah Bay and around Cape Flattery looking for locations to build lifesaving stations, and finally to the scene of the wreck itself off of Vancouver Island.[4] Their report to the president, including conclusions and recommendations, was published on April 14, 1906.

As noted elsewhere, the commission was careful not to assign blame too readily or to treat the officers and men too harshly. The report noted, for instance, that had either or both shore parties hiked to and remained at the top of the cliff near where the *Valencia* had wrecked, there would have been people to retrieve the lines that were in fact fired from the ship; if there had been people there to secure those lines, many fewer people would have died. However, the commission did not "desire to attribute blame to this shore party for its failure to grasp this opportunity." A group of tired, frightened people, some injured and all cold and wet from many hours of exposure, could not—the commission felt—be expected to make the same sorts of intelligent decisions made by men not under the same sort of duress.

As for the Canadian participants in the abortive rescue attempts, the commission said simply, "the Commission does not deem it proper to criticise [*sic*] the conduct of other than American citizens."[5] Thus, the Canadians—while many accusations were leveled at them from within their own country—were not subject to the same level of examination and inquiry as were the Americans involved. The US commission refused to comment on the behavior (or misbehavior) of, for instance, the men aboard the *Czar* and the *Salvor*, two rescue vessels whose captains—Christensen and Troupe, respectively—decided that there was no one left on board *Valencia* to save and therefore simply steamed away, in spite of sailors aboard *Queen* noting that there *were* in fact, people left alive on deck and in the rigging of the stricken vessel.[6]

Of course, the presence of coastal rescue services or more ships might not have done any good. Clay Evans, former superintendent of the Canadian Coast Guard at Bamfield, notes, "We have to keep in mind that even if there was a coastal lifeboat stationed within a short distance of the disaster, they would have had a phenomenally difficult time recovering hypothermic, barely alive civilians from a ship breaking up in the surf zone."[7] That could just as easily be applied to lifesaving efforts launched from *Queen* or another vessel. It could be that, even with all the chatter, the accusations of murder, and complaints about the putative rescuers not doing more, the *Valencia* was doomed the moment she struck the rocks and no rescue would have been possible, at least not from ships at sea.

Still, others were not so reluctant to criticize. A Norwegian whaling captain who was on board the whaler *Orion* when it ran close to the reef to see if anyone was left alive denounced Captain Cousins, Captain Patterson, and those on board the ships that responded to the wreck as "being guilty of nothing less than cold-blooded murder" for not moving in closer to the ship or not sending lifeboats to the wreck.[8] Survivor Frank Bunker used almost the same language when speaking to a *Seattle Times* reporter on January 27, characterizing the deaths as "wholesale murder."[9]

"Murder" seems a bit strong, but many observers felt that, as a description of what befell the passengers and crew of *Valencia*, it wasn't far off. Recounting the *Czar* leaving the site of the wreck, one source calls it, not without reason, "the most shameful incident in Canadian maritime history."[10] Neither the rescue vessels nor the official US and Canadian lifesaving services conducted themselves with the valor, dedication, and effectiveness we've come to expect from our mariners *or* our official rescue organizations. As the commission's report noted, "there was certainly no display of the heroic daring that has often marked other such emergencies in our merchant marine."[11]

One *Valencia* survivor, Cornelius Allison, a crusty seaman with many years' experience on the water, testified during the earlier Seattle hearings that "the surf that finally battered the Valencia to pieces could not be called high" and noted that "the vessels at sea stood off and made no attempt to lower a boat. . . . There might have been some excuse for the *Queen*'s not coming in closer [because the water was too shallow for the larger ship], but there was a tug lying alongside of her that did not come any closer than the large vessel. It all looked wrong to us."[12]

One of the men rescued from a life raft was passenger George Harraden. Though more circumspect than some other witnesses, carefully avoiding the use of such inflammatory terms as "murder," he did testify that he thought the tug that approached *Valencia*, *Czar*, could have gone in closer. If she had done so, he said, she could have floated a line in or

This photo shows a lifeboat from *City of Topeka* rescuing passengers in one of *Valencia*'s life rafts. Notice that, even partially unloaded, the raft is almost swamped, and this in fairly calm water.

drifted one of *Queen*'s rafts over to the stranded ship. He admitted that the effort might not have been successful, but he insisted that it should have been attempted.[13]

As an early book about the history of Seattle notes, "There is no doubt that the *Queen* made almost no effort and the *Topeka* but little, to rescue the men and women clinging half-frozen to the rigging of the doomed vessel."[14]

There was plenty of blame to go around. The tule life jackets may have been less useful than desired. The shore parties should have known that the crew and passengers would be best served if they had stayed at the top of the bluff rather than heading off into the forest in a risky and time-consuming search for help. The captains and crews of the rescue vessels might have, and perhaps should have, braved the breakers and rocks and deployed lifeboats in an attempt to save those on board *Valencia*. As the commission's report notes, "It was practically the unanimous opinion of a large number of witnesses that the ordinary lifeboats could have been safely taken in toward the wreck as long as they kept outside of the line of breakers. Outside of this line the sea was not combing or breaking, and small boats would have been perfectly safe."[15] More ships, small tugs, especially, should have been brought to the scene. Certainly, the rescue vessels ought not to have given up—or been sent away—before every last chance of saving lives had been exhausted. The blood of the passengers and crew of the wrecked ship was—and remains—on many hands.

But the commission charged with investigating the *Valencia* incident knew where the majority of the blame lay. And as circumspect as they were, and though they were determined to be as fair as possible, they did not hesitate to place that blame squarely where it belonged. In the end, they singled out two ultimate causes.

First, they blamed a large part of the debacle on "the defective state of the aids to navigation and preservation of life in the shape of light-houses, fog signals, life saving [*sic*] equipment, and means of communication in the vicinity of the wreck." That is, the commission found fault with—and laid a large portion of the blame on—the deplorable state of lifesaving infrastructure near the Strait of Juan de Fuca, a situation that had contributed to the deaths of many hundreds of people over the years, and one in which both US and Canadian officials were complicit. In particular, the commission noted that the "most important light in this entire course, to wit, that on Cape Flattery, is not placed in the zone where, by reason of the fog and the thick weather, the greatest danger lies."[16] In an area known long before the *Valencia* incident as "the graveyard of the Pacific," they seemed to be asking, why was more not done to prevent further loss of life?

In the main, though, the report laid the blame—regretfully, because he could not be present to defend himself—squarely at the feet of *Valencia*'s captain, Oscar M. Johnson, noting that Johnson "appears to have been a man of good character, sober, and with a good reputation as a seaman," but describing his management of the *Valencia* as "unsatisfactory

The so-called graveyard of the Pacific extends roughly from just north of Tillamook Bay, Oregon, to the northwestern tip of Vancouver Island.

on several points."[17] The report then enumerates those points, the most important of which appear to be as follows:

1. Johnson allowed his lookouts to remain on station much longer than the two or so hours recommended for the conditions; the result was most likely an incapacitated, or at least overly tired and understandably inattentive, lookout on the bow when the ship struck the rocks.

2. He did not require boat drills during the voyage. Given that as many as half of the crew were men new to this vessel, this omission "nullified to a large extent the usefulness of the boat equipment so far as this trip was concerned."[18] (It also meant that the *passengers* were also largely unfamiliar with lifeboat and abandon ship procedures.)

3. Johnson insisted that his log was "overrunning" by about 6 percent, in spite of the contrary opinion of some of his officers. This was coupled with his lack of understanding "that at this time of year the normal [that is, the Davidson] current flows toward the northward."[19]

4. Although he saw no land or lights after passing Cape Mendocino at 5:30 a.m. on Sunday, the captain "did not commence to take soundings until 6 p.m. Monday, thirty-six hours later, when his last definite point of departure was at least 450 miles behind him."[20]

5. When he did take soundings, he did not take them with sufficient frequency, nor did he interpret them correctly. As the report points out, the soundings that he did take "might not have shown him where he was, but if properly studied they would at least have demonstrated the fact that he was not where he thought he was and that he should be on his guard."[21]

6. Given the doubt as to his position, Johnson should have exercised more prudence and headed out to sea until his position could be ascertained.

In the end, the commission's findings can be boiled down to one painful declaration. After considering the evidence and the testimony of passengers and crew, the investigators found that "upon [Johnson's] improper navigation . . . must rest the primary responsibility for the disaster."[22]

Chapter 32

The Men of the Commission

These days, when we think of government investigating committees and commissions, we tend to think of large deliberative bodies—often official Senate or House committees or subcommittees—holding public hearings under the glare of lights and the scrutiny of cameras and media representatives. The Watergate committee, more properly known as the Select Committee on Presidential Campaign Activities, for instance, appointed by the US Senate in 1973 to investigate the Watergate scandal, consisted of seven men, aided by two main counsels and ten staff members. (One of the latter *was* female, Jill Wine-Banks, an attorney who would eventually become the first woman to serve as US General Counsel of the Army. Wine-Banks was the only female officially associated with the investigating committee.) The hearings were broadcast live every day and replayed at night, for a total of 319 hours of airtime.[1]

Similarly, the 1965 Illinois Crime Investigating Commission looking into charges of corruption in the Illinois General Assembly was made up of eight investigators, again, all male and was aided by multiple other supporters and secretaries.

In 1948, the US Senate established the lyrically named Permanent Subcommittee on Investigations of the Governmental Affairs Committee to investigate issues of waste and fraud throughout the executive branch. That committee—which still exists—has nine members.

The investigation of the *Valencia* wreck was, compared to such undertakings, small scale. The commission consisted of only three members, and none were senators or congressmen; they were merely government

In 1977, after her Watergate experience, Jill Wine-Banks (then Jill Wine-Volner) became general counsel of the Army.

workers, although two of them were relatively high-ranking federal employees. Recall, though, that they served at the specific request of the president of the United States and answered directly to him; the commission was composed of only three people, but there was real power behind it.

The members of the small but impressively named Federal Commission of Investigation were Lawrence O. Murray, assistant secretary of commerce and labor, who headed the commission; Herbert Knox Smith, deputy commissioner of corporations; and Capt. William Turnbull Burwell, of the US Navy. The three were empowered to call witnesses but could not compel them to testify. Likewise, the commission could make recommendations, but could not enforce them. They could, however, forward those recommendations to President Theodore Roosevelt, who could in turn forward them to agencies that *could* ensure that the recommendations were acted upon.[2]

Of the three men, only one was not a political functionary. William Turnbull Burwell was an 1866 graduate of the US Naval Academy and a former inspector of ordnance and commander of, among other vessels, the USS *Wheeling*, a gunboat and convoy escort, and the USS *Alexander*, a transport that became a collier later in its life.[3] Born in 1846 in Vicksburg, Mississippi, by 1906, when the *Valencia* went down, he was commandant of the Navy Yard in Puget Sound. He was well acquainted with the area and with the so-called graveyard of the Pacific. Burwell would pass away in 1910, at the age of sixty-three, having attained the rank of rear admiral.[4]

Herbert Knox Smith was, in 1906, deputy commissioner of corporations, a businessman and a close friend of President Theodore Roosevelt.[5] A graduate of Yale Law School, Smith would run unsuccessfully for governor of Connecticut in 1912. He was a major in the US Army during World War I and worked as a lawyer after the war. As deputy commissioner of corporations, Smith was part of the Bureau of Corporations, the predecessor to today's Federal Trade Commission.[6] He would die in 1931.[7]

Lawrence O. Murray as a younger man

William Turner Burwell after his appointment as rear admiral

The chair of the commission was assistant secretary of commerce and labor Lawrence O. Murray. Born in 1864 in rural Pennsylvania, Murray studied law and was admitted to the bar in New York. In 1908, Murray would become comptroller of the currency, having previously worked for the Treasury Department and acted as secretary of the Central Trust company. Murray was a political powerhouse, a member of Roosevelt's famous "tennis cabinet," a group of younger men who provided informal support and advice to the president. After his retirement, Murray spent a great deal of time in Africa, following his passion for big game hunting. He passed away in 1926 in Elmira, New York.[8]

It was a small group, seemingly diverse, but in some ways very much alike. Two of the men were essentially politicians, friends of the powerful governmental elite and members of the elite themselves. Used to hobnobbing with presidents and high-level government and banking officials, Smith and Murray were two of a kind; both were capable of being circumspect and tactful and were well aware of the niceties to be

Deputy Commissioner of Corporations Herbert Knox Smith

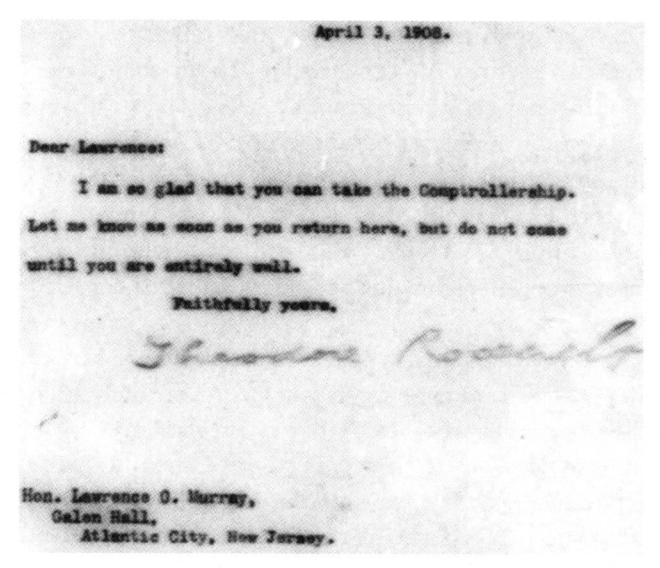

April 3, 1908.

Dear Lawrence:

I am so glad that you can take the Comptrollership. Let me know as soon as you return here, but do not come until you are entirely well.

Faithfully yours,

Theodore Roosevelt

Hon. Lawrence O. Murray,
Galen Hall,
Atlantic City, New Jersey.

After the *Valencia* investigation, President Theodore Roosevelt offered Murray the position of comptroller of the currency. This is Roosevelt's letter back to Murray after the latter accepted the appointment.

observed when dealing with the powerful, but they were also comfortable interacting with ordinary people. Nonetheless, they wielded a great deal of power and were used to getting their way; when they spoke, they expected to be listened to. When they asked a question, they were—as attorneys, as senior managers, and as representatives of the United States of America—used to a prompt and germane response.

At first, Burwell strikes us as a different animal. Not familiar with or necessarily comfortable in the halls of political power, Burwell was a naval officer, not a politician—a military man, a servant of the people, not a power broker or a dealmaker. He is cut from a different cloth. Still, Burwell was comfortable working within a military hierarchy, and he had ascended (and would ascend further) within that hierarchy to a position of a good deal of responsibility and power. Like the others, Burwell too was used to having his orders obeyed and his questions answered.

Collectively, the three men were a formidable force. Two were attorneys; they were used to digging for the truth and were not afraid—and may even have relished the opportunity—to argue or debate in order to arrive at that truth. Burwell, though not a politician, was nonetheless used to wielding power; at the time, he had risen to the command of a large naval base, supply depot, and shipyard, with hundreds of men reporting to him, either directly or through intermediaries. He had commanded powerful and sophisticated naval vessels. He was a man of the sea, which the other two men were not, and he must have commanded the respect of those men for his specialized nautical knowledge, experience, and training.

These were not men with which to trifle. They were used to digging for—and ultimately finding—the facts, and most likely were also used to dealing with people who attempted to prevaricate, to obscure those facts, or to slant or shade their recollection of them. One of them was intimately familiar with the ways of the sea and of men who sailed it. If one were testifying in front of this commission, it would be wise to simply tell the truth, as best as one could recall it. As near as can be told from reading through the report issued by the commission, the trio seems to have accomplished their goal of determining who was at fault in the *Valencia* disaster and what should be done about it.

Chapter 33

Lives Cut Short

As with all such tragedies, there are, embedded in the larger narrative, heartrending tales of lives—some of them quite young—cut short. One such story was that of young Iva Shaver and her fiancé, Bert Parker.[1] Traveling from Los Angeles and bound for Seattle to marry and to take advantage of the area's then-cheap land, the couple was traveling with Iva's mother, Mrs. W. C. Rosenberg. (One assumes that Iva's mother—whose surname is sometimes given as Rosenberger—accompanied the as-yet-unmarried the pair as a chaperone, in the name of propriety.) The two would never marry, of course; all three died in the wreck.

There was another young man from Los Angeles on board, Jim McCarrick, but we know next to nothing about him. McCarrick was not on any official passenger lists; he was, in many books and articles, listed as a stowaway, but it's more likely that he had signed on at the last minute as a cook. It was rumored that he had been accused of a crime in the Pacific Northwest and was making his way back to Seattle in order to prove his innocence.[2] If so, he never got that chance.

Another couple, the T. J. (or F. J.) Campbells, formerly of Alameda, California, were also traveling to Seattle. Their sixteen-year-old stepdaughter, Edith, was accompanying them. The elder Campbell was an insurance agent but had resigned his position in order to travel to Seattle to start a machine business with an acquaintance. (Some sources, for example, Hooper, have "F. J." as the husband's initials, and list his first name as Frank. That may in fact be more likely; it would certainly be no great surprise for "F. J." scribbled in a reporter's notebook to be mistaken for "T. J.," only to be corrected years later by other researchers.[3]) Mr. Campbell would survive, becoming part of the Bunker party that traveled overland looking for help, but would lose both his wife and his stepdaughter in the wreck.[4]

Somewhat eerily, it is said that one of the officers' wives, a woman married to fourth mate Herman Aberg, received a predeparture visit from a fortune teller who predicted that Herman would be shipwrecked, leaving her a widow. Herman himself laughed at the dire prophecy and left on the journey to Seattle. He died in the water off of Vancouver Island, along with so many others.[5] While many of the bodies were never recovered, and many of those that were could not be identified, Aberg's body was one of those that was identified, mainly because he wore a distinctive blue sweater and a monogrammed ring.

Many of the survivors—not that there were all that many; only thirty-seven people lived through the sinking—disappeared from public view afterward, going on to lead quiet lives, no doubt grateful for having lived but scarred by all they had seen and endured. We do know something about a few of them, though.

John Segalos (sometimes rendered as Joe Cigalos), the courageous Greek fireman who dove into the freezing water to try to swim a line to shore, not only survived but was awarded several medals for his bravery and was said by newspapers of the time to have been endorsed as a Carnegie Hero Fund beneficiary, though that may be untrue.[6] Rescued from a life raft, along with several others, by the *City of Topeka*, Segalos made a decent living touring after the wreck, billed as "The Hero of the Valencia," performing swimming and diving exhibitions across the country, recounting the disaster, and showing off his cherished medals. Segalos was later assaulted and robbed, losing his cherished medals in the robbery. He died impoverished at the age of seventy-six.

In addition to the two shore parties already described, there were twenty-three survivors in two life rafts, one of which was eventually picked up by the *City of Topeka*. That raft had nineteen men on it, overloading it by at least one, and was partially swamped when *Topeka* spotted it. (Fireman John Segalos was on that raft.) The men on the raft barely survived, having spent several hours in the wet, cold raft. Another raft started off with ten men aboard, but six of them succumbed to the cold during their fourteen hours on the raft; the remaining men on that raft were found alive on a small island off the coast of Vancouver Island.

By far the survivor about whom we know the most is Frank Forest Bunker. He was, after all, not just an articulate and oft-quoted survivor of the *Valencia* incident, he was also a well-known educator and a published

A *Valencia* life raft, possibly one of those picked up by rescue vessels. Notice its low freeboard; this is essentially just a floating platform, easily swamped in seas of any size.

researcher and philosopher of education. (To a great extent, it is Bunker who popularized what became known as the "junior high school movement." Prior to his involvement, and as recounted in his 1935 book, *The Junior High School Movement—Its Beginnings*, there was really no such thing as a junior high school in the United States; students went straight from an elementary school of one sort or another into high school.[7] Bunker is one of the key figures in the movement, seen as quite progressive at the time, which was meant to provide a transitional step during which students could grow and mature physically and emotionally, as they studied basic subjects the mastery of which would help them in their high school and college lives.)

Though he remarried not long after the wreck, Bunker remained childless, perhaps never quite recovering from the loss of his young son and daughter.[8] He continued his professional ascent, receiving his PhD in 1913, afterward working for the United States Bureau of Education, the federal department that would eventually become the US Department of Education. He eventually became an editor for the Carnegie Institution of Washington and wrote several books relating to education and to his travels in the South Seas, most of which he undertook on behalf of the Bureau of Education.

Some of those travels resulted in a dark spot that looms over Bunker's career—at least, in retrospect—and which may negatively influence our judgment of him. Generally regarded as a forward-looking educator, Bunker was nonetheless undeniably a product of his time: In the 1919 *Federal Survey of Education*, he participated in what was touted as a scientific study aimed at guiding educational reform in what was then the territory of Hawaii. Unfortunately, that survey was used to implement some fairly hurtful xenophobic colonial policies over the Nikkei (people of Japanese ancestry), the territory's largest ethnic group. Bunker supported, for instance, the idea of prohibiting the Nikkei from learning Japanese at school, citing various World War I–era prohibitions on foreign language teaching, such as Nebraska's Siman Act, a 1919 ban on German language instruction.

Admittedly, this was a popular position at the time. In a country scarred by the then-recent war, paranoia about "foreigners" ran (and to

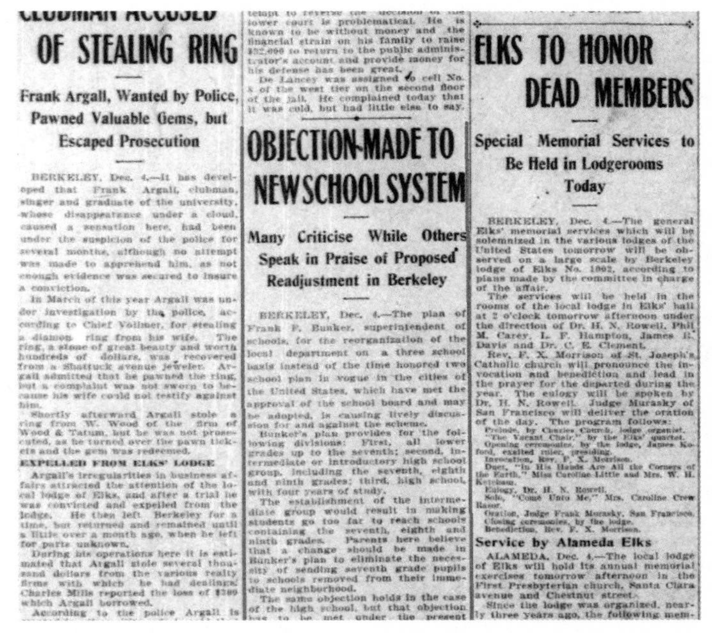

CLUBMAN ACCUSED OF STEALING RING

Frank Argall, Wanted by Police, Pawned Valuable Gems, but Escaped Prosecution

BERKELEY, Dec. 4.—It has developed that Frank Argall, clubman, singer and graduate of the university, whose disappearance under a cloud, caused a sensation here, had been under the suspicion of the police for several months, although no attempt was made to apprehend him, as not enough evidence was secured to insure a conviction.

In March of this year Argall was under investigation by the police, according to Chief Vollmer, for stealing a diamond ring from his wife. The ring, a stone of great beauty and worth hundreds of dollars, was recovered from a Shattuck avenue jeweler. Argall admitted that he pawned the ring, but a complaint was not sworn to because his wife could not testify against him.

Shortly afterward Argall stole a ring from W. Wood of the firm of Wood & Tatum, but he was not prosecuted, as he turned over the pawn tickets and the gem was redeemed.

EXPELLED FROM ELKS' LODGE

Argall's irregularities in business affairs attracted the attention of the local lodge of Elks, and after a trial he was convicted and expelled from the lodge. He then left Berkeley for a time, but returned and remained until a little over a month ago, when he left for parts unknown.

During his operations here it is estimated that Argall stole several thousand dollars from the various realty firms with which he had dealings. Charles Mills reported the loss of $300 which Argall borrowed.

According to the police Argall is

...tempt to reverse the decision of the lower court is problematical. He is known to be without money and the financial strain on his family to raise $32,000 to return to the public administrator's account and provide money for his defense has been great.

De Lancey was assigned to cell No. 8 of the west tier on the second floor of the jail. He complained today that it was cold, but had little else to say.

OBJECTION MADE TO NEW SCHOOL SYSTEM

Many Criticise While Others Speak in Praise of Proposed Readjustment in Berkeley

BERKELEY, Dec. 4.—The plan of Frank F. Bunker, superintendent of schools, for the reorganization of the local department on a three school basis instead of the time honored two school plan in vogue in the cities of the United States, which have met the approval of the school board and may be adopted, is causing lively discussion for and against the scheme.

Bunker's plan provides for the following divisions: First, all lower grades up to the seventh; second, intermediate or introductory high school group, including the seventh, eighth and ninth grades; third, high school, with four years of study.

The establishment of the intermediate group would result in making students go too far to reach schools containing the seventh, eighth and ninth grades. Parents here believe that a change should be made in Bunker's plan to eliminate the necessity of sending seventh grade pupils to schools removed from their immediate neighborhood.

The same objection holds in the case of the high school, but that objection has to be met under the present

ELKS TO HONOR DEAD MEMBERS

Special Memorial Services to Be Held in Lodgerooms Today

BERKELEY, Dec. 4.—The general Elks' memorial services which will be solemnized in the various lodges of the United States tomorrow will be observed on a large scale by Berkeley lodge of Elks No. 1002, according to plans made by the committee in charge of the affair.

The services will be held in the rooms of the local lodge in Elks' hall at 2 o'clock tomorrow afternoon under the direction of Dr. H. N. Rowell, Phil M. Carey, L. F. Hampton, James R. Davis and Dr. C. E. Clement.

Rev. F. X. Morrison of St. Joseph's Catholic church will pronounce the invocation and benediction and lead in the prayer for the departed during the year. The eulogy will be spoken by Dr. H. N. Rowell. Judge Murasky of San Francisco will deliver the oration of the day. The program follows:

Prelude, by Charles Church, lodge organist.
"The Vacant Chair," by the Elks' quartet.
Opening ceremonies, by the lodge, James Koford, exalted ruler, presiding.
Invocation, Rev. F. X. Morrison.
Duet, "In His Hands Are All the Corners of the Earth," Miss Caroline Little and Mrs. W. H. Ketcham.
Eulogy, Dr. H. N. Rowell.
Solo, "Come Unto Me," Mrs. Caroline Crew Bazor.
Oration, Judge Frank Murasky, San Francisco.
Closing ceremonies, by the lodge.
Benediction, Rev. F. X. Morrison.

Service by Alameda Elks

ALAMEDA, Dec. 4.—The local lodge of Elks will hold its annual memorial exercises tomorrow afternoon in the First Presbyterian church, Santa Clara avenue and Chestnut street.

Since the lodge was organized, nearly three years ago, the following mem-

Frank Bunker's call to establish a junior high school system truly was revolutionary and raised something of a furor, as reported in this *San Francisco Call* article on December 5th, 1909.

some extent continues to run) rampant; perhaps we should not allow his involvement in such intolerant goings-on to cloud our judgment at this late date.

In any case, Bunker led a long and productive life, appearing as late as 1941 on CBS radio programs devoted to education and faith. He passed away on September 25, 1944, in Suffolk, Massachusetts, at the age of seventy-one, and was buried in Washington, DC.

Epilogue

In the hills above Seattle, Washington, some forty-five miles from the Strait of Juan de Fuca, sits the Mt. Pleasant Cemetery. Situated in the northern portion of the Queen Anne Hill section of the city, the cemetery is—as most cemeteries are—quiet, peaceful, and beautiful, much of it shaded by old maple and chestnut trees that have been growing there for well over one hundred years. (The earliest burials there occurred in the late 1870s, and the grounds have been expanded and beautified many times since those days.)

It is a sad place, of course, but a beautiful spot to rest, to pause and consider one's mortality, and to reflect upon the lives and deaths of the people who are buried on the grounds.

The mass grave marker at Mt. Pleasant Cemetery

A great many moderately famous people are interred at Mt. Pleasant: Filipino author Carlos Bulosan; Anna Herr Clise, cofounder of Children's Orthopedic Hospital; the magnificently named Orange Jacobs, a superior court judge who was for a time also a mayor of Seattle; and African American leader and founder of the Black sorority Delta Sigma Theta Bertha Pitts Campbell, among several others.[1]

But one of the most meaningful places at Mt. Pleasant is a gravesite that bears no name. Instead passersby see only a timeworn inscription on a weathered gray marker:

Dedicated to the Unknown Dead
of the
Valencia Disaster
Jan. 22, 1906
Off Vancouver Island

Memorial Services
Sept. 23, 1906

Remains Brought And
Interred By
Organized Labor

There is no one famous buried beneath the marker. In fact, we don't really know *who* lies there. The mass grave contains the remains of twenty-six unknown and unidentifiable victims of the *Valencia* tragedy, gathered from the beaches around Vancouver Island or found floating in the waters nearby. The stark, colorless stone stands as testament to the memory of those who died. "Let's not forget them," the monument seems to say, "Keep them in our thoughts, nameless though they might be."

The predominant impact of the *Valencia* story, the thing that ultimately strikes most people, is the utter unfairness, the sheer injustice of it: This is a tale in which innocent people drowned within sight of the shore, including women and children who perished even as they viewed what should have been—and what they must have thought would certainly be—their salvation. And all so *needlessly*; these were blameless victims of accident, of laziness, of caprice, of haughty arrogance and misplaced pride. Of the many, many errors made by the captain, the shore parties, and the putative rescuers, it's obvious looking back at the incident that if any one or two or three of those fateful mistakes had been avoided, people would not have died in such number. In the end, hubris killed all of those people and destroyed the ship to which they entrusted their lives.

Valencia had a rough life. She was old and tired, and she had collided with other vessels, been lost, been grounded and beaten upon by storms, and—finally—been crushed by the wind and the waves against the unforgiving rocks of Walla Walla Reef. Some, pointing to her history of maritime mishaps, have said that she was an unlucky ship, for sailors are a superstitious lot. The reality, though, is that there is no such thing as an unlucky ship: there are poorly built or maintained ships, and there are unskilled captains. And there is the wild, inconstant, implacable sea, relentless and eager to take advantage of the slightest lapse. She was neither an unlucky ship nor a lucky one; she was just a ship. And as sometimes happens with ships, she was mishandled by men who did not know the ocean as well as they thought, and so she was swallowed by the sea, and most of her crew and passengers died.

The sad reality is that, in 1906, there were simply no lifesaving services available on *either* side of the strait. In the many wrecks that occurred off the coast of Vancouver Island before 1906, those that did survive did so *in spite* of their absence and because of the presence of the trail, the telegraph, and the lineman's huts that were scattered along the trail. It was not enough, but it was a beginning.

We like to believe that we have learned lessons from the sinking of the SS *Valencia*, as we have learned from the sinking of the *Titanic*—two disasters caused in large part by the hubris of the men who captained the two ships. But have we *really* learned those lessons? Do we know enough to guarantee that the hubris, the excessive pride, of some men and women will not continue to cause disasters such as these? Perhaps not. One has only to look at the *Challenger* disaster to understand that, quite often, people can convince themselves that an undertaking is safe simply because they are in positions of power and *wish* it to be so. But as others have said, "hope is not a strategy."

The ship herself has no actual epitaph, but perhaps a fitting one might be found in the words of science fiction writer Lois McMaster Bujold:

> The dead cannot cry out for justice.
> It is a duty of the living to do so for them.[2]

This book is an attempt to help tell the story of the *Valencia* and the fate of her passengers and crew. It's a story that deserves to be better known, and one that the author was surprised to find few people, even those living on Vancouver Island, have heard. But if we can remind the public of the *Valencia* tragedy, perhaps the victims can rest more peacefully—as can we, having done our best to do our duty by all of the lost souls that perished there.

Acknowledgments

To my wife, Lesley, who read various iterations of this manuscript, and was fearless enough to offer constructive criticisms and helpful suggestions—thank you for sticking with me all these years, through thick and thin. Especially the thin. There was a lot of thin.

My editor at Lyons Press/Sheridan House, Brittany Stoner, was a supportive and reassuring writing partner and a valuable sounding board as we worked our way through the story and how best to tell it.

Early readers and contributors included CDR. John Harrington, USCG (Ret.), a longtime long-distance sailor; Clay Evans, retired superintendent of the Bamfield Station, Canadian Coast Guard, and an early and articulate critic of Canada's response to the *Valencia* incident as well as its pre-*Valencia* preparations; Dr. Herb Benavent, rigging instructor, author, and long-distance sailor; geological engineer Robert Mahood, of Okotoks, Alberta, Canada, who offered particularly useful and insightful suggestions about the manuscript and who was kind enough to help search the Canadian archives; Rick Brown, longtime friend and a respected advisor on all things related to mathematics and physics; Heather Feeney, exhibits manager for the Maritime Museum of British Columbia, for her time and expertise, and for the opportunity to photograph the few extant *Valencia* artifacts; Molly Dumas, for her incredible illustrations that added so much to the book; Dr. Richard Thompson, oceanographer, for his help in understanding how ocean currents form and their effects on navigation; and Robert Mester, of Northwest Maritime Consultants, for his time and his expertise relating to the murky and convoluted subject of maritime salvage.

As always, any remaining errors, omissions, or ambiguities are mine alone.

It's always wonderful to get suggestions and comments from fellow writers. Patricia Wood (author of *Lottery* and *Cupidity*, and a long-distance sailor herself) and Val Davisson (author of the Logan McKenna mystery series) are two of my favorites. Thank you for your help and advice over the years and, in particular, for reading and commenting on early drafts of *Ship of Lost Souls*.

Finally, and perhaps most of all, thank you to the readers. Despite poorly researched Internet memes asserting that fewer and fewer people read, it turns out that many millions of you *do* read and have shown that you're hungry for new information, a great story, an escape, an uplifting word. Books, albeit often in new and unforeseen forms, are still incredibly popular, and human beings of all stripes are voracious consumers of novels, history, how-to, romance, sci-fi, and other genres. Books, in other words, are still important, as are painting, drama, music, photography, sculpture, and all the many forms of art that bring us together, teach us about the world, and warm our souls. Readers like you warm the souls of authors everywhere. Thank you for warming mine.

Rod Scher
Depoe Bay, Oregon
January 2024

Notes

Prologue

1. John MacFarlane, in "The Valencia," The Nauticapedia, 2017, https://www.nauticapedia.ca/Gallery/Valencia.php, disputes some aspects of the account by Michael C. Neitzel in the latter's book, *Final Voyage of the Valencia:* Amazing Stories (S.l.: Heritage House Publishing, 2020), MacFarlane maintains that it was his grandfather, George Alexander MacFarlane, who discovered the lifeboat in a field in 1933 and that when he did, the boat was filled with scrap metal parts. According to that version of the story, the elder MacFarlane removed the boat's nameplate and stored it in a cupboard for many years. Of course, both stories could be true, with the boat appearing in the bay and then, shortly after its appearance, being dragged ashore and the nameplate being subsequently removed. Note, though, that MacFarlane says that 117 people lost their lives in the incident; while we do not know *exactly* how many people died, we do know that it was more than 117. In any case, the nameplate from the lifeboat now resides in the collection of the Maritime Museum of British Columbia.

Chapter 1

1. "Professor Bunker in Trouble," *Press Democrat*, September 16, 1902.

Chapter 2

1. "Cramp Shipbuilding, Philadelphia Pa.," William Cramp & Sons Shipbuilders, accessed May 21, 2023, http://shipbuildinghistory.com/shipyards/large/cramp.htm. The company went out of business in 1927 but was temporarily brought back to life in the 1940s with an infusion of $22 million from the US Navy. In all, Cramp and Sons would produce a total of 576 vessels, including seven submarines and one cruiser.

2. Soumya Chakraborty, "Designing a Ship's Bottom Structure—A General Overview," Marine Insight, July 14, 2022, https://www.marineinsight.com/naval-architecture/design-of-ships-bottom-structure/.

3. Ibid.

4. Ibid.

5. Lawrence O. Murray, *Wreck of the Steamer Valencia. Report to the President, of the Federal Commission of Investigation, April 14, 1906* (Washington: Govt. Print. Off., 1906).

6. Daryl C. McClary, "Valencia, SS, the Wreck of (1906)," July 29, 2005, https://www.historylink.org/File/7382.

7. Neitzel, *Final Voyage of the Valencia*. It turns out that even *stainless steel* can be corroded by sugarcane juices. Studies have compared the effects of sugarcane juice on various types of steel because, after all, the cane is processed in machines *made* of steel. See https://tinyurl.com/yuyay72v.

8. Murray, *Wreck of the Steamer Valencia*.

9. "What of the Bulkheads of the S.S. Valencia?" *The Seattle Star*, January 30, 1906.

10. McClary, "Valencia, SS, the Wreck of (1906)."

11. "S/S Edam (2), Holland America Line," Edam (2), Holland America Line, accessed May 24, 2023, http://www.norwayheritage.com/p_ship.asp?sh=edama.

12. "S/S La Normandie, C.G.T.—Compagnie Générale Transatlantique (French Line)," Norway Heritage, accessed May 24, 2023, http://www.norwayheritage.com/p_ship.asp?sh=lanoe.

13. Neitzel, *Final Voyage of the Valencia*.

14. Henry G. Preble, *Chronological History of the Origin and Development of Steam Navigation* (Philadelphia: L. R. Hamersly & Co., 1895).

Chapter 3

1. "Eugène Édouard Désiré Branly," Institute of Chemistry, April 2002, http://chem.ch.huji.ac.il/~eugeniik/history/branly.html.

2. Ibid.

3. Carole E. Scott, "The History of the Radio Industry in the United States to 1940," EHnet, accessed May 4, 2023, https://eh.net/encyclopedia/the-history-of-the-radio-industry-in-the-united-states-to-1940/.

4. The SS *St. Paul* was a 15,150-ton transatlantic liner (also built by Cramp and Sons) that typically sailed from Southampton to New York via Cherbourg, France. Like *Valencia*, she was drafted (actually, contracted) into service during the Spanish-American War.

5. Interestingly, the *Reina Mercedes*, built around the same time as the *Valencia*, boasted what *Valencia* sorely lacked: multiple watertight compartments meant help resist flooding. (See https://www.spanamwar.com/reinam.htm.)

Chapter 4

1. More than 265 people were killed and fifty-four were injured when the *Maine*'s forward magazine exploded. The explosion was, rightly or wrongly, used to inflame passions and generate support for a declaration of war against Spain. President McKinley's declaration of war was dated April 21 and was approved by Congress on April 25. The Spanish, meanwhile, declared war on April 23.

2. Patrick McSherry, "Why Was the Spanish-American War Fought?" The Spanish-American War Centennial Website, November 29, 2021, https://www.spanamwar.com/why.html. Of course, "Remember the Maine!" was a popular rallying cry, referencing the US Navy vessel that exploded in Havana Harbor. The explosion—the

cause of which was never fully explained—accelerated the declaration of war, as perhaps it was meant to, according to some theories.

3. "The Spanish-American War," TopSCHOLAR Database, accessed December 19, 2023, https://digitalcommons.wku.edu/wtw_that_little_war/. (Also see www.archives.gov/publications/prologue/1998/spring/spanish-american-war.)

4. These days, the use of charters continues, and the military often charters aircraft, as well as vessels, buses, and other forms of transportation.

5. Patrick McSherry, "The U.S. Army Transport Service," The Transport Service—Army Transport Ships During the Spanish American War, accessed December 17, 2023, https://www.spanamwar.com/transports.htm. Much of the information here comes from this website.

6. *Cramp's Shipyard: The William Cramp & Sons Ship and Engine Building Company, 1830* (Philadelphia: The Company, 1910), see 3 and 157. Most sources indicate that the original USS *Maine* was built in the New York Naval Shipyard.

7. Then again, it was a war, a disaster by any measure, and almost by definition.

8. This was not a trivial issue. The major killer of soldiers during the war was not combat but typhoid fever, which is spread through the contamination of food or water by Salmonella typhi bacteria; this in turn is often caused by untreated sewage.

9. McSherry, "The U.S. Army Transport Service."

Chapter 5

1. "USCG Timeline 1700s–1800s," United States Coast Guard (USCG) Historian's Office, accessed May 13, 2023, https://www.history.uscg.mil/Complete-Time-Line/Time-Line-1700-1800/.

2. Ibid.

3. Ibid. Interestingly, in 2018, *another* vessel named *Priscilla*, this one loaded with fertilizer bound for England from Lithuania, also ran aground, this time off the coast of Scotland.

4. "A Heavy Sea Running," National Archives and Records Administration, 1987, https://www.archives.gov/publications/prologue/1987/winter/us-life-saving-service-1.html.

5. "The Long Blue Line: 'You Have to Go Out, but You Don't Have to Come Back'—Origin of the Old Life-Saving Service Motto," United States Coast Guard #62, My Coast Guard News, April 8, 2022, https://www.mycg.uscg.mil/News/Article/2978215/the-long-blue-line-you-have-to-go-out-but-you-dont-have-to-come-backorigin-of-t/.

6. "U.S. Coast Guard Missions: A Historical Timeline," US Coast Guard Missions: A Historical Timeline, accessed June 15, 2023, https://media.defense.gov/2021/Jun/04/2002735330/-1/-1/0/USCGMISSIONSTIMELINE.PDF.

7. James D. Charlet, "Island History: The 1876 Fatal Disaster Mystery of the Nuova Ottavia," *Island Free Press*, March 21, 2022.

8. Ralph C. Shanks, Wick York, and Lisa Woo Shanks, *The U.S. Life-Saving Service: Heroes, Rescues, and Architecture of the Early Coast Guard* (Novato, CA: Costaño Books, 1996).

9. Ibid.

10. Geraldine McCaughrean, Victor G. Ambrus, and Herman Melville, *Moby Dick* (Oxford: Oxford University Press, 2016).

11. Shanks, York, and Shanks, *The U.S. Life-Saving Service*. Some sources have said that waves on the Pacific coast have been known to tower up to two hundred feet high, though former Canadian Coast Guard Superintendent Clay Evans said in an email (January 22, 2024) to the author that he thinks that's unlikely.

Chapter 6

1. Joe Palca, "Alabama Woman Stuck in NYC Traffic in 1902 Invented the Windshield Wiper," NPR, July 25, 2017, https://www.npr.org/2017/07/25/536835744/alabama-woman-stuck-in-nyc-traffic-in-1902-invented-the-windshield-wiper#.

2. *Wireless at Sea: The First 50 Years* (Chelmsford, England: Marconi International Marine Communication Company, Limited, 1950). The *Lake Champlain* was a 446-foot steel steamer built in Scotland. The ship's Marconi equipment was fitted into a specially constructed windowless cabin measuring less than sixteen square feet.

3. Ibid.

4. Len Buckwalter, *ABC's of Short-Wave Listening* (Indianapolis: H. W. Sams, 1970).

5. *Wireless at Sea.*

6. Ibid.

7. Ibid.

Chapter 7

1. Robert C. Belyk, *Great Shipwrecks of the Pacific Coast* (New York: Wiley, 2001). A "knot" is a unit indicating a vessel's speed relative to the water through which it is traveling. It is equal to 1.852 kilometers per hour or approximately 1.15 miles per hour. One describes a vessel's speed as "xxx knots," *not* as "xxx knots per hour." Saying that a boat was going "27 knots per hour" is a sure way to get laughed out of a marina, chandlery, or other nautical environment.

2. John Frazier Henry, "The Wreck of the Valencia," *Columbia: The Magazine of Northwest History*, 1993.

3. Some sources, including some then-contemporary newspaper articles, have the vessel's name as *Queen City*, rather than *Queen.* This is most likely incorrect, as most authoritative sources, including the report issued by the official investigating committee and many other contemporary references, have it as *Queen*. In fact, in the annals of the Pacific Steamship Company archives themselves, one sees many references to a company vessel named *Queen*, and none to *Queen City.* See https://tinyurl.com/mr28frby and https://tinyurl.com/4md3226a. The same is true when viewing the Online Archive of California's "Guide to the Pacific Coast Steamship Companies Collection." (Some contemporary newspaper accounts, such as one in the *San Francisco Call* 83, no. 94 [March 4, 1898]: 3, do have the ship correctly as *Queen*.) In addition, the 1905 *Reports of the Department of Commerce and Labor* note that it was *Queen*—rather than *Queen City*—that sustained fire damage in January 1904, recording that fourteen people died in or subsequent to the blaze.

4. Neitzel, *Final Voyage of the Valencia.*

5. Tyler Hooper, "The Titanic of the Pacific," *The Atavist Magazine*, May 2, 2023, https://magazine.atavist.com/the-titanic-of-the-pacific-valencia-shipwreck-disaster-british-columbia/. Hooper's in-depth and well-researched article has *Queen* correctly as one of the rescue vessels rather than misnaming her *Queen City*, as do some other sources.

6. Neitzel, *Final Voyage of the Valencia.*

7. Ibid.

8. K. E. Harlow, "Shipwrecks in British Columbia's Waters," harlowmarine.com, May 11, 2015, https://www.harlowmarine.com/shipwrecks-in-british-columbias-waters/. Also see "Shipwrecks of the West Coast Trail," Hikewct.Com, West Coast Trail Guide, accessed September 12, 2023, https://hikewct.com/index.php/shipwrecks.

Chapter 8

1. "The Great 1906 San Francisco Earthquake," US Geological Survey, accessed December 31, 2023, https://earthquake.usgs.gov/earthquakes/events/1906calif/18april/.

2. Murray, *Wreck of the Steamer Valencia*. The official government tally notes sixty-five crewmembers and 108 passengers, seventeen of whom were women. It also notes that there were "a few children."

3. Gordon Newell and Joe Williamson, *Pacific Coastal Liners* (Seattle, WA: Superior Pub., 1959). Some passengers were actually scheduled to go on to Juneau, Alaska, after stopping at Seattle and Vancouver.

4. Neitzel, *Final Voyage of the Valencia*. As Neitzel points out, it's difficult to see how Johnson could have served the steamship company for fifteen years because he lived in McKeesport, in south-central Philadelphia, for several years.

5. McClary, "Valencia, SS, the Wreck of (1906)."

6. Ibid. Neitzel, whose research is generally solid but who may be mistaken here, has him as Petterson; see the bibliography. In Neitzel's defense, "Peterson" may simply be an Americanization of the Swedish "Petterson." Both names simply mean "son of Peter," after all.

7. Murray, *Wreck of the Steamer Valencia.*

8. Some early news accounts confused bosun McCarthy with J. McCarthy, a messboy who perished in the wreck.

9. David H. Grover, *The Unforgiving Coast: Maritime Disasters of the Pacific Northwest* (Corvallis: Oregon State University Press, 2002). Note that McCarthy's testimony was given, in this case, not to the American committee investigating the disaster but before the coroner's jury in Victoria, British Columbia.

10. Hooper, "The Titanic of the Pacific."

11. Ibid. Whether a fireman or a messboy, he was at least officially part of *Valencia*'s crew and should therefore have outranked Frank Bunker, a mere—though important—passenger.

Chapter 9

1. Ibid.

2. "Forecast for January 20," *San Francisco Call*, January 20, 1906, sec. 1.

3. Hooper, "The Titanic of the Pacific."

4. Murray, *Wreck of the Steamer Valencia*.

5. Fog and wind are not usually present at the same time, but it can happen. There are different types of fog. Radiation fog occurs when solar energy exits the earth and the temperature matches the dew point. This type of fog generally dissipates as the temperature rises during daylight. Strong wind is unlikely with radiation fog. Advection fog, on the other hand, occurs when air that is warmer than the ground moves over the ground surface. Warm, moist air blows in (often from the south) and comes into contact with cooler moisture on the ground—or the surface of the ocean. Its movement means, almost by definition, that wind will be present. (See https://tinyurl.com/577zvsca.)

6. In many recountings, Holmes's first name is unknown, and the mate is simply listed as "W." But he is listed as "William" in "West Coast Trail Shipwrecks," Hikewct.com, accessed August 4, 2023, https://hikewct.com/index.php/shipwrecks/valencia/1valencia/246-thelost.

7. Neitzel, *Final Voyage of the Valencia*.

8. Murray, *Wreck of the Steamer Valencia*.

9. Ibid. We have no way of knowing whether Captain Johnson wrestled with the possibility of heading back out to sea to wait out the storm and then approaching again in daylight. He certainly *should* have considered it, but there has been no recorded testimony regarding any conversations with his crew about doing so. If he indeed considered it and couldn't make up his mind to head to sea, then his indecision cost many lives, including his own.

10. "SS Humboldt," Skagway Stories, February 22, 2013, https://www.skagwaystories.org/2013/02/22/ss-humboldt/. The captain of the *Edith* at that time was most likely Thomas A. Miller because it was Miller who was in command a year or so later when *Edith* received a radio call to sail to Mouat Point, North Pender Island, just east of Vancouver Island, to rescue passengers from the stranded SS *Humboldt*.

Chapter 10

1. Adam Augustyn, "Seven Wonders of the World," Encyclopædia Britannica, accessed September 24, 2023, https://www.britannica.com/topic/Seven-Wonders-of-the-World.

2. William S. Hanable, "Lightships on Washington's Outer Coast," Historylink.Org, February 3, 2005, https://www.historylink.org/File/7189.

3. Clay Evans, "All but Forgotten: Early Measures for Maritime Safety on Canada's West Coast," *The Northern Mariner / Le Marin Du Nord*, December 2021, 387–408, https://doi.org/10.25071/2561-5467.55.

4. Ibid.

5. Ibid.

Chapter 11

1. "Cape Flattery Lighthouse," Lighthouse Friends, accessed October 10, 2023, https://www.lighthousefriends.com/light.asp?ID=120. A blockhouse is essentially a small fort, built in such a way as to allow defenders to fire in multiple directions.

2. A Fresnel (pronounced "fray-nel") lens consists of a series of concentric rings that act to focus a beam of light. This sort of lens, invented in the 1820s by French physicist Augustin-Jean Fresnel, is much lighter than an equally effective lens made of a solid glass disk would be.

3. "Cape Flattery Lighthouse," United States Coast Guard, accessed October 10, 2023, https://www.history.uscg.mil/Browse-by-Topic/Assets/Land/All/Article/2015165/cape-flattery-lighthouse/.

4. "Cape Flattery Lighthouse," Lighthouse Friends.

5. Ibid.

6. Ibid.

7. T. Wheeler, "John M. Cowan," Lighthouse Research Catalog, November 9, 2018, https://archives.uslhs.org/people/john-m-cowan.

8. "Cape Flattery Lighthouse," Hikewct.com, accessed October 10, 2023, https://hikewct.com/index.php/atoz-glossary/cape-flattery-light.

9. Chiefly used as foghorns, a diaphone is a device that produces a blast of two long, deep, powerful tones that most often end with a sort of "grunt." (There is also a linguistic definition that is not relevant here.)

10. Murray, *Wreck of the Steamer Valencia*. In fact, as noted on page 47 of the commission's report, neither the Cape Flattery Light nor the lighthouse at Carmanah experienced fog that night, so neither activated its foghorn.

Chapter 12

1. A strait is a natural narrow waterway that connects two larger bodies of water. A canal is much like a strait, but the former is manmade.

2. You *can* drive across from Qualicum Beach to Tofino on Highway 4, but that only transits the lower third of the island and requires driving a somewhat roundabout route totaling over 101 miles. Farther north, you can also drive (on Highway 28) from Campbell River on the east coast inland, but only to Gold River, a bit more than halfway across—and a drive of some fifty-four miles.

3. Kendra Wong, "Hundreds of Tiny Tremors Recorded Between Victoria and Seattle," Victoria News, June 11, 2018, https://www.timescolonist.com/local-news/hundreds-of-tiny-tremors-recorded-between-victoria-and-seattle-4670827. People tend to think of—and refer to—Vancouver Island as running north and south and think of the area where Valencia struck the reef as the west coast of the island. (On the east coast are population centers such as Ladysmith and Nanaimo. The west coast is still much less populated than the east.) Some confusion can arise, though, when one looks at a map of the area: Vancouver Island is not really oriented north and south; instead it's tilted such that the true orientation of the "northern tip" is almost exactly northwest, while its "southern tip"—say, where Victoria lies—lies almost exactly to the southeast. Thus,

referring to the coasts as east and west is largely a convenient fiction and can confuse the unwary, at least at first.

4. "Vancouver Island," New World Encyclopedia, accessed October 2, 2023, https://www.newworldencyclopedia.org/entry/Vancouver_Island.

5. The 1846 Oregon Treaty, mainly meant to decide the question of Oregon's borders, also awarded Vancouver Island to the British.

6. Jeff Wallenfeldt, "Vancouver Island," Encyclopædia Britannica, October 1, 2023, https://www.britannica.com/place/Vancouver-Island.

7. "Killing the Indian in the Child," Facing History & Ourselves, September 20, 2019, https://www.facinghistory.org/en-ca/resource-library/killing-indian-child. Some of the Canadian residential schools went as far as concealing children's deaths, burying hundreds in unmarked graves. This is not ancient history; the last of those schools did not close until 1998.

8. "Vancouver Island," New World Encyclopedia.

9. Note that Vancouver, to his credit, was known to enjoy good relations with Indigenous people he encountered during his travels in the Pacific Northwest and elsewhere.

10. The Spanish impact on the island is sometimes forgotten, but the Spanish have been exploring and trading on Vancouver Island since 1774, when they sent a frigate, *Santiago*, there to establish a presence. Later, the Spanish built Fort San Miguel near Yuquot in 1789. Ironically, the Strait of Juan de Fuca itself was named not by (or for) a Spanish explorer but by British fur trader Charles W. Barkley, who named it after a *Greek* navigator sailing on a Spanish expedition in 1592.

11. September 30 is observed as Orange Shirt Day in Canada: orange shirts with the slogan "Every Child Matters" are sold and the color is often incorporated into signage and social media posts.

Chapter 13

1. "The Atlantic Daily News," *Atlantic Daily News: Hamburg* (1906), accessed August 9, 2023, https://earlyradiohistory.us/1906hamb.htm. "Steerage" is an interesting term referring to those passengers housed deep in the bowels of the ship, through which ran the ropes and cables that controlled the ship's rudder. That is, the ropes and cables that were used to "steer" the vessel.

2. "Wireless Ship Act," *Annual Report of the Commissioner of Navigation, Department of Commerce and Labor, Bureau of Navigation*, 1911, 43–56.

3. Sharon L. Morrison, "Radio Act of 1912," 2009, https://www.mtsu.edu/first-amendment/article/1090/radio-act-of-1912.

4. George F. Worts, *Modern Electrics* 3, no. 6 (September 1910).

Chapter 14

1. Tyler R. Grindstaff, *Wreck of the Steamship Valencia*, vol. 2, 2 vols. (Grindstaff Publishing, 2020). The two-volume Grindstaff book is not really a narrative describing the incident; instead it is the collected transcripts of testimony given to the officials who

investigated the *Valencia* tragedy. That said, it is an extremely useful tool as a chronicle of events and lends much insight into various perspectives on those events.

2. Note that the power of those waves breaking over the beleaguered ship would have been impossible to resist. One cubic meter (about thirty-five cubic feet) of salt water weighs roughly 1,000 kilograms (about 2,200 pounds). A large wave striking a vessel would tear it apart and would instantly sweep an unprepared person off the ship and into the cold, turbulent water.

3. Grindstaff, *Wreck of the Steamship Valencia*, vol. 2.

CHAPTER 15

1. McClary, "Valencia, SS, the Wreck of (1906)."

2. Murray, *Wreck of the Steamer Valencia*.

3. Scott Yorko, "The West Coast Trail Is a Beautiful Hike with a Horrifying History," *Backpacker*, October 28, 2022, https://www.backpacker.com/trips/adventure-travel/canada/the-west-coast-trail-is-a-beautiful-hike-with-a-horrifying-history/.

4. Murray, *Wreck of the Steamer Valencia*.

5. Some sources have the name as Richley.

CHAPTER 16

1. Clay Evans, email to Rod Scher, Valencia, July 25, 2023.

2. Murray, *Wreck of the Steamer Valencia*.

3. Ibid.

4. Ibid. *Valencia*'s last inspection, which she passed with flying colors, was on January 6, 1906.

5. Newell and Williamson, *Pacific Coastal Liners*. A few decades ago, most life preservers were made with kapok, a fiber obtained from the seed pods of the kapok tree, which is native to tropical Asia, but today most use a buoyant plastic foam of some sort inside.

6. Neitzel, *Final Voyage of the Valencia*.

7. "Lifebelts Sink When Wet," *The Morning Oregonian*, January 30, 1906. Mind you, this was from a particularly negative, and somewhat sensationalistic, article in the *Morning Oregonian*. The article's subheads include such examples of journalistic objectivity as "Steamer Queen Ignores Calls for Help," "*Valencia* Manned by an Incompetent Crew," and "Discipline Is Lost When Struggle for Boats Begins." All of which may be true, but which do not reflect the objectivity we've come to expect—or at least hope for—in modern news sources. Whitney would resign his position in 1917 to become a surveyor for the American Bureau of Shipping, but there was no talk at the time of any sort of misconduct. In his previous position, he had conducted several investigations of sinkings, both before and after the Valencia incident.

CHAPTER 17

1. McKenna Ehrmantraut, "The Evolution of Women's Swimming: From Then to Now," Swimming World News, *Swimming World Magazine*, October 20, 2022, https:

//www.swimmingworldmagazine.com/news/the-evolution-of-womens-swimming-from-then-to-now/.

2. Jenny Landreth, "How Women Took the Plunge in Britain's Waters" inews.co.uk, July 16, 2020, https://inews.co.uk/inews-lifestyle/votes-for-swimming-the-feminist-history-of-taking-the-plunge-72024. In fact, until 1901, men and women who congregated on beaches in the United Kingdom were actually segregated, and female swimmers were not even allowed to participate in the Olympics until 1920.

3. Murray, *Wreck of the Steamer Valencia*.

4. Keep in mind that the terms "left" and "right" are used somewhat loosely here, given that the island itself is situated NW/SE.

Chapter 18

1. Neitzel, *Final Voyage of the Valencia*. In addition to Cadet Willits, there were actually five full-fledged US Navy members on board *Valencia* at the time of the sinking, three holding the rank of ordinary seaman and two classed as coal passers; all five drowned.

2. Murray, *Wreck of the Steamer Valencia*.

3. After a public outcry subsequent to the wreck of the *Valencia*, the trail would be widened, improved, and maintained as an official lifesaving trail. Eventually, it would become part of Canada's national parks' West Coast Trail.

4. "West Coast Trail: Challenges," Pacific Rim National Park Reserve, November 19, 2022, https://parks.canada.ca/pn-np/bc/pacificrim/activ/SCO-WCT/i.

5. "The Valencia Disaster," Hikewct.com, accessed September 4, 2023, https://hikewct.com/index.php/shipwrecks/valencia/5thebunkerparty.

6. The party seems not to have considered what seems a sensible middle ground: a few of the men heading off to find help while leaving *part* of the group behind to watch for a line fired from the ship. On the other hand, former Canadian Coast Guard Superintendent Clay Evans is of the opinion that the men on board *Valencia* were probably not particularly well trained in the use of the Lyle gun, a tool that's usually used to send a line from the shore *to* a ship, and one with which coastguardsmen practiced and drill regularly.

Chapter 19

1. Yes, that bit of nautical terminology is the source of our various sayings having to do with "sticking things out 'til the bitter end," and other such adages. (And it is also the ultimate source of the name of a still popular New York City nightclub.)

2. "David Lyle and His Life-Saving Gun," National Parks Service, accessed May 18, 2023, https://www.nps.gov/spar/learn/historyculture/david-lyle.htm.

3. "Sumner Increase Kimball," United States Coast Guard—Notable People, accessed May 13, 2023, https://www.history.uscg.mil/Browse-by-Topic/Notable-People/All/Article/1762434/sumner-increase-kimball/.

4. Lyle was an incredibly productive man with varied interests. Retiring from the US Army as a colonel, he wrote scientific papers on ornithology and geology and invented manufacturing processes related to leather, files, and other materials. He died on October 11, 1937.

5. Grover, *The Unforgiving Coast.*
6. Murray, *Wreck of the Steamer Valencia.*
7. "The Valencia Disaster: The McCarthy Party," Hikewct.com, accessed September 4, 2023, https://hikewct.com/index.php/shipwrecks/valencia/1valencia/238-themccarthy.

Chapter 20

1. Murray, *Wreck of the Steamer Valencia.*
2. Ibid.
3. "The Valencia Disaster: The McCarthy Party." Nonetheless, this decision does fly in the face of the party's stated purpose, which was to land as close to the wreck as possible and then climb the bluff so that a line could be fired to them from the stricken ship. It was an understandable failure, but a failure regardless, and one of many. McCarthy later told the *Seattle Star* that the underbrush was so dense "that it was simply beyond human strength to force one's way through the tough tangle."
4. Susie Quinn, "Meet Minnie Paterson, the Heroine of Cape Beale," *The Vancouver Island Free Daily*, May 2, 2021.

Chapter 21

1. Neitzel, *Final Voyage of the Valencia.*
2. "Investigation Turns on Discipline of Crew, Value of Life Belts, and Other Topics," *San Diego Union and Daily Bee*, February 2, 1906.
3. Grover, *The Unforgiving Coast.*
4. Henry, "The Wreck of the Valencia." The full headline read: "Second Officer Patterson [*sic*] of the Valencia Says Life Boats Could Have Reached Stricken Vessel Any Time During Tuesday or Wednesday—Investigation Now Being Carried On." (Headlines were not particularly brief in those days.)
5. Murray, *Wreck of the Steamer Valencia.*
6. "Master of Vessel Says He Was in Full Command," *The Los Angeles Herald*, February 23, 1906.
7. Murray, *Wreck of the Steamer Valencia.*
8. Ibid. Mind you, Clay Evans, former Canadian Coast Guard superintendent, points out in an email to the author (January 22, 2024) that tugboats *also* draw a lot of water; thus, a tug may not have been able to get in much closer to the wreck than the *Queen* or other large vessels. Perhaps they could have gotten in a little closer and then drifted in life rafts, assuming they had them on board.
9. Neitzel, *Final Voyage of the Valencia.*
10. Ibid.
11. "West Coast Shipwrecks: The Valencia," Hikewct.com, accessed September 4, 2023, https://hikewct.com/index.php/books-wct-shipwrecks/book-valencia. The author of this piece pulls few punches, calling this portion of Neitzel's book "a trainwreck of a chapter."
12. Neitzel, *Final Voyage of the Valencia.*
13. Ultimately, *Topeka* would rescue nineteen people, all of whom were in an overloaded life raft that was slowly drifting out to sea.

14. Murray, *Wreck of the Steamer Valencia*. Neitzel criticizes the three men, believing that they should have made an attempt to rescue the survivors, perhaps by tying a rock to a line and throwing it to the people on board; after all, they were only yards from the ship and situated about one hundred feet above the sea, so they had a good angle from which to attempt a throw. The criticism is difficult to answer, although they—and the people on board—would have had to have worked *very* quickly if they were to save anyone.

CHAPTER 22

1. All of the following temperatures are indicated in degrees Fahrenheit.
2. "Cold Shock," Cold Water Safety, accessed December 26, 2023, https://www.coldwatersafety.org/cold-shock.
3. We're not speaking here of a gentle gasp, such as might result when one is surprised by someone seeing another unexpectedly coming around a corner in front of them. This is a *deep* involuntary inhalation in which one's lungs suddenly and completely fill with air—or, if submerged, with water. Cold-water shock can also cause heart attacks in individuals with heart problems; naturally, this could also cause drowning. Keep in mind that one can drown after inhaling only a half a pint (about nine ounces) of water. (See https://rnli.org/safety/know-the-risks/cold-water-shock.)
4. "Cold Shock."
5. Rick Curtis, "Outdoor Action Guide to Hypothermia #38; Cold Weather Injuries," Princeton University, accessed December 26, 2023, https://www.princeton.edu/~oa/safety/hypocold.shtml.
6. "Hypothermia," The Mayo Clinic, March 5, 2022, https://www.mayoclinic.org/diseases-conditions/hypothermia/symptoms-causes/syc-20352682.
7. "Cold Shock."
8. "Sea Water Temperature in British Columbia in January," SeaTemperature.info, accessed December 26, 2023, https://seatemperature.info/january/british-columbia-water-temperature.html.
9. Remember that, as discussed in chapter 17, few aboard the ship were likely to be good swimmers, and that's especially true of women and children.
10. Gene Little, "U.S. Coast Guard Auxiliary—District 9er," USCG Auxiliary, February 1, 2016, https://wow.uscgaux.info/content.php?unit=092&category=cold-water. In fact, the USCG Auxiliary defines "cold" water as anything less than 69 degrees F.

CHAPTER 23

1. Interview with Bob Mester, of Northwest Maritime Consultants LLC, by Rod Scher, October 27, 2023. The earliest known salvage laws, "the Rules of Oleron," were written prior to 1180 AD in France. These ancient laws are sometimes known as the "Lawes of Pleron."
2. R. S. Pattox, "Coastwise Navigation on the Pacific," *Pacific Marine Review*, n.d., 66–66.
3. George L. Canfield, George W. Dalzell, and J. Y. Brinton, *The Law of The Sea: A Manual of The Principles of Admiralty Law for Students, Mariners, and Ship Operators*

(D. Appleton & Co., 1921). According to Canfield et al., the percentage of the value awarded to a salvor generally depends on seven things: the degree of danger from which the lives or property are rescued; the value of the property saved; the risk incurred by the salvors; the value of the property employed by the salvors in the wrecking enterprise, and the danger to which it was exposed; the skill shown in rendering the service; the time and labor occupied; and the degree of success achieved, and the proportions of value lost and saved.

4. The issue of shipwrecks, and thus potential salvage questions, is still very much with us, as Bob Mester points out in an email to the author dated January 14, 2024: "Today with all our advances in engineering, navigation, communications and emergency responses we still have events like the SS *El Faro* in 2006. She was lost at sea with her entire crew of 33 on October 1, 2015, after steaming into the eyewall of Hurricane Joaquin."

5. Interview with Bob Mester, of Northwest Maritime Consultants LLC, by Rod Scher, October 27, 2023. Mester himself was a member of a US Marines special operations force before going into commercial salvage.

6. Clay Evans, email to Rod Scher, January 22, 2024.

Chapter 24

1. Cory Mitchell, "Big Blue: Nickname for IBM, Overview, History," Investopedia, May 18, 2023, https://www.investopedia.com/terms/b/big-blue.asp. According to IBM's 2022 *Annual Report*, the company generated over $60 billion in revenue during that year, so the company is doing well, even without its personal PC line, which it sold to Lenovo in 2005. We're using two examples of American companies, but there are many companies around the world that are similarly enduring. Nintendo, which started out in 1889 manufacturing playing cards, is still going strong. The Olde Bell, a hotel and pub located in Hurley, Berkshire, England, on the bank of the River Thames, was originally opened as a hostelry in the year 1135, so it's nearly one thousand years old. A brewery founded by Belgian monks, the Affligem Brewery, has been in continuous operation since the year 1074. (Some companies, on the other hand, were quite ephemeral: Elizabeth Holmes's blood-testing company, Theranos, existed only from 2013 to 2018. Then again, that company was built on fraudulent claims that led the founder to prison, where she still resides. Blockbuster Video, which failed to keep up with innovations in technology, mainly the advent of streaming video, was another relatively short-lived company; it was founded in 1985 and filed for bankruptcy in 2010.)

2. Gerald M. Best, *Ships and Narrow Gauge Rails: The Story of the Pacific Coast Company* (San Diego, CA: Howell-North, 1964). Unless otherwise noted, most of the information in these paragraphs comes from Best.

3. Timelines are necessarily somewhat fuzzy here. After the 1916 merger with the Admiral Line, the holding company went into liquidation in 1925, but the corporation existed as an entity until 1925. The provenance of corporations, as of people, is not always entirely clear. There were several instances of PCSSC mergers occurring after rate wars, with some sources noting that the companies got together to "stabilize" the rates as a way of ending the pricing conflicts, which were, after all, injurious to the companies' bottom lines. Today such attempts at "stabilization" would most likely be labeled "collusion."

4. Murray, *Wreck of the Steamer Valencia.*

5. "Pacific Coast Steamship Company (1877–1916)," Islapedia, October 30, 2022, https://www.islapedia.com/index.php?title=Pacific_Coast_Steamship_Company_1877-1916.

6. PCSSC books such as *All About Alaska* (first published in 1888) were essentially long PR pieces extolling the virtues—real and imagined—of the places visited by PCSSC vessels.

7. Best, *Ships and Narrow Gauge Rails.*

8. Not much has changed, really. People still generally assume—correctly or not—that those who supposedly wield political (or other) power are often in fact manipulated, or at least influenced by, the wealth and pressure brought to bear by powerful people in business.

9. "Lifebelts Sink When Wet."

Chapter 25

1. Richard J. Goodrich, "The Titanic of the West," *Medium*, November 12, 2022, https://medium.com/lessons-from-history/the-titanic-of-the-west-bdd0e83e5ca3.

Chapter 26

1. Nathaniel Philbrick, *In the Heart of the Sea: The Tragedy of the Whaleship* Essex (New York: Penguin Books, 2001).

2. There is, in fact, an entire branch of scientific study known as *currentology*.

3. In the Southern Hemisphere, they are—not surprisingly—forced instead to the left.

4. Dr. Richard Thomson, email to Rod Scher, Davidson Current, July 31, 2023.

5. Ibid.

6. Jörn Callies et al., "Seasonality in Submesoscale Turbulence," *Nature Communications* 6, no. 1 (2015).

7. According to Royal Museums Greenwich, one statute mile is equal to about 0.868 nautical miles. See https://www.rmg.co.uk/stories/topics/nautical-mile.

Chapter 27

1. After many months (or even years) at sea, a gam could take on a party air, with sailors shouting greetings, trading mail and news, sharing food and drink, and more. Occasionally, the master of one ship might board the other vessel to visit with his fellow captain. For men cooped up on board a small vessel, it was great fun to see and speak with someone new. (In chapter 53 of *Moby Dick*, Ahab refuses to board the *Albatross*, a ship that the *Pequod* has encountered, and cuts their gam short; this is one way we come to realize that Ahab's monomaniacal quest for the white whale has begun to outweigh both his common sense and his humanity.)

2. An excellent description of the lives of seamen on an early hiding voyage can be found in Richard Henry Dana's *Two Years Before the Mast*.

3. Evans, "All but Forgotten."

4. Martin Hollmann, "Radar: Christian Huelsmeyer, The Inventor," Radar, 2007, https://www.radarworld.org/huelsmeyer.html.

5. Radio direction finding, or RDF, *was* in use in the early 1900s, but it wasn't until 1910 and later that it was commonly found on vessels and in lighthouses. RDF is another technology that could have aided *Valencia* as she approached the coast of Vancouver Island.

Chapter 28

1. Evans, "All but Forgotten."

2. "Pachena Point Lighthouse," Lighthouse Friends, accessed August 26, 2023, https://www.lighthousefriends.com/light.asp?ID=1200.

3. Around this time, and largely as a result of the sinking of the RMS *Titanic*, a frenzy of maritime safety-related efforts occurred, including the planning and building of many aids to navigation and the adoption of the first International Convention for the Safety of Life at Sea (SOLAS).

4. Donald Graham, *Keepers of the Light: A History of British Columbia's Lighthouses and Their Keepers* (Madeira Park, BC: Harbour Pub. Co., 1996).

5. Christopher Cole and John M. MacFarlane, "The Lifeboats at Bamfield BC," The Nauticapedia, 2008, https://www.nauticapedia.ca/Gallery/Bamfield_Lifeboats.php. After the *Valencia* disaster, the Bamfield station would be the recipient of the world's first purpose-built *powered* coastal lifeboat. Ironically, the boat was actually made in America and originally intended for the fledgling US Life-Saving Service, but was purchased by the Canadian Department of Marine and Fisheries. A similar boat was placed at Neah Bay.

6. Evans, "All but Forgotten." Evans not only served with the Canadian Coast Guard for over thirty-five years but was for many years the commanding officer of the Bamfield Lifeboat Station.

7. Cole and MacFarlane, "The Lifeboats at Bamfield BC."

8. "New Coast Guard Station Breaks Ground on Vancouver Island," Vancouver Island CTV News, May 28, 2019, https://vancouverisland.ctvnews.ca/new-coast-guard-station-breaks-ground-on-vancouver-island-1.4440913.

9. Sequim, in spite of its spelling, is pronounced "Skwim."

10. Greg Bradsher, "Assignment: Neah Bay, Washington, 1909; the United States Revenue-Cutter Service and the USRC Snohomish," National Archives and Records Administration, September 1, 2020, https://text-message.blogs.archives.gov/2020/09/01/assignment-neah-bay-washington-1909-the-united-states-revenue-cutter-service-and-the-usrc-snohomish/.

11. Ibid.

12. Harlow, "Shipwrecks in British Columbia's Waters."

13. Cameron La Follette, "Santo Cristo de Burgos," *The Oregon Encyclopedia*, July 12, 2022, https://www.oregonencyclopedia.org/articles/manila-galleon-wreck-on-the-oregon-coast/#.YzT9pC2B1QI. Beeswax was valuable for a number of reasons, including the fact that it made wonderful, long-lasting (and thus very expensive) candles that burned clean and with a pleasing scent. In particular, the Catholic church prized candles

made of beeswax, seeing a rich and detailed symbolism in both its pure light and its pale wax, the latter of which they associated with Christ's flesh. Thus, beeswax candles were (and still are) preferred for mass and other Church functions. (See https://www.catholic.com/magazine/print-edition/mind-your-beeswax.)

Chapter 29

1. "West Coast Trail: Packing," Pacific Rim National Park Reserve, November 19, 2022, https://parks.canada.ca/pn-np/bc/pacificrim/activ/SCO-WCT/iv. It might be useful to note here that the trail's official seventy-five-kilometer length is reckoned to be wildly optimistic by many; some say that the trail is at least fifteen kilometers (about ten miles) longer than that. Also, when looking up recommendations for Canada's Pacific Rim Park and its West Coast Trail, do not confuse it with the Pacific Crest Trail, which runs through California, Oregon, and Washington. (For example, dogs are allowed on many parts of the latter, but not on the former.)

2. "The West Coast Trail," Hikewct.com, accessed November 11, 2023, https://www.hikewct.com/index.php/component/tags/tag/ladders.

3. Jason Hummel, "The West Coast Trail: Vancouver Island's Iconic Hike," Switchback Travel, accessed September 1, 2023, https://www.switchbacktravel.com/west-coast-trail.

4. "West Coast Trail History: Upnit Lodge: Bamfield Accommodation," Upnit Lodge, May 8, 2019, https://upnitlodge.ca/west-coast-trail-history/.

5. In Victoria, British Columbia's 34th Annual Report (1913), several improvements to the trail were listed, including building out the trail from Banfield Creek for sixteen miles as a wagon road, and for the balance, of approximately sixteen miles, as a foot trail. In addition, the report notes the building of "five shelter huts . . . between Nootka and Quatsino." See https://open.library.ubc.ca/media/download/pdf/bcbooks/1.0222220/0.

6. "Parks & Trails," Vancouver Island News, Events, Travel, Accommodation, Adventure, Vacations, January 13, 2020, http://vancouverisland.com/things-to-do-and-see/parks-and-trails/.

7. Maxwell W. Finkelstein, "Pacific Rim National Park Reserve," *The Canadian Encyclopedia*, December 20, 2006, https://www.thecanadianencyclopedia.ca/en/article/pacific-rim-national-park-reserve.

8. Neitzel, *Final Voyage of the Valencia*. Neitzel is one who believes that the desertion, especially by the Bunker party, was a cowardly act that doomed the remaining survivors.

9. Ibid., 41. In the end, the two parties reported the wreck to authorities within an hour or two of one another, between 1 p.m. and 3 p.m., on Tuesday, January 23.

10. At the bluff overlooking the spot where *Valencia* went down, at kilometer eighteen of the West Coast Trail, are two of Parks Canada's famous red Adirondack chairs. (Parks Canada has a history of placing two such chairs at spots that are either especially pleasant to view or especially meaningful.)

11. Murray, *Wreck of the Steamer Valencia*.

Chapter 30

1. Grover, *The Unforgiving Coast*.

2. Ibid.

3. Ibid. Metcalf was a career politician who had attended Yale College and then Yale Law School and was eventually admitted to the Connecticut and New Yorks bars. President Roosevelt appointed him Secretary of Commerce and Labor in July 1904. In late 1906, he was appointed Secretary of the Navy.

4. Ibid.

5. Evans, "All but Forgotten."

6. Ibid.

7. The (Canadian) Report of the Coroner's Inquest *does* exist but doesn't tell us much about the causes of the wreck itself, only how some of the victims died, for example, trauma, hypothermia, drowning, etc. See *Office of The Coroner, Coroner's Inquisition into the Sinking of the SS Valencia* § (n.d.). Dominion of Canada, 1906.

8. Clay Evans, Zoom interview by Rod Scher, November 29, 2023. In addition to being the officer-in-charge at the Bamfield Coast Guard (Lifeboat) Station, Evans was for quite a long time also on the board of the Maritime Museum of British Columbia where, as it happens, several remnants of the *Valencia* reside.

9. Evans, "All but Forgotten."

10. Graham, *Keepers of the Light.*

11. Evans, "All but Forgotten." While the government of Canada dithered, more people died: in 1906–1907, for example, the bark *Skagit* went down near Clo-oose, and the bark *Coloma* foundered off of Cape Beale.

Chapter 31

1. Humphrey was actually born in Indiana and practiced law there in the 1880s and 1890s. He moved to Seattle in 1893. After a long run as a congressman, Humphrey was, in 1925, appointed a member of the FTC until President Franklin D. Roosevelt fired him in 1933. (Humphrey sued Roosevelt and eventually won, with the Court determining that the then-president had exceeded the limits of his presidential power.)

2. Grover, *The Unforgiving Coast.*

3. Bradsher, "Assignment."

4. *Columbine*, built in 1892, was a 155-foot vessel with a beam of about twenty-six feet and a draft of just over fifteen feet. Smaller than *Queen*, she could get in fairly close to the wreck site and to other areas that were being considered as potential sites for future lighthouses and lifesaving stations. *Columbine* had an impressive history; she was eventually sold in 1927 but remained in service under various owners until 1942, when she was abandoned.

5. Murray, *Wreck of the Steamer Valencia.*

6. Note that *Salvor* had no direct contact with *Valencia*. She was told to turn back by *Czar*, after *Czar*'s captain reported that there were no living souls aboard the larger ship.

7. Clay Evans, email to Rod Scher, January 22, 2024.

8. Henry, "The Wreck of the Valencia."

9. Ibid. Bunker's editorial comment was posed in the form of questions, ending with: "It might be well to ask why the Pacific Coast Steamship Company is willing to permit the possibility of such conditions on their boats and why the United States

Government [*sic*] does nothing to prevent such wholesale murder of her citizens." (See the *Seattle Daily Times* of January 27, 1906.) Mincing words was not something of which Bunker was often accused.

10. Neitzel, *Final Voyage of the Valencia*.

11. Murray, *Wreck of the Steamer Valencia*.

12. "West Coast Trail Shipwrecks."

13. Ibid.

14. Clarence Bagley, *History of Seattle from the Earliest Settlement to the Present Time* (Chicago, IL: The S. J. Clarke Publishing Company, 1916). Keep in mind, though, that *Topeka* remained on station and did in fact rescue a large number of men from *Valencia*'s life rafts.

15. Murray, *Wreck of the Steamer Valencia*.

16. Ibid.

17. Ibid. Note too that, on the day of the sinking, Capt. J. B. Patterson, aboard the *Topeka*, told a newspaper reporter from the *San Francisco Examiner* that Johnson was "a good navigator—a first-class man. . . . I regard him as a capable man."

18. Ibid.

19. Ibid. In fact, the commission characterized Johnson's navigation as "haphazard," overall.

20. Ibid.

21. Ibid.

22. Ibid.

Chapter 32

1. "Senate Watergate Panel Begins Today—Inquiry on Alleged Campaign Sabotage," *New York Times*, May 17, 1973.

2. It's important to keep in mind that the commission certainly had no power to compel testimony (or action) from the Canadian participants in the *Valencia* incident. In the end, the commission simply said that it would be improper for it to "criticise [*sic*] the conduct of other than American citizens" and essentially ignored the Canadian contributions to the incident and its attendant rescue efforts.

3. In nautical usage, a collier is a ship that transports coal used to refuel other vessels.

4. "Captain Burwell Promoted," *Colusa Daily Sun* XXVIV, no. 133 (June 6, 1906). Seemingly as a reward for his service and for his contributions to the investigating commission, Burwell was promoted to rear admiral shortly after the commission turned in its report and disbanded.

5. Smith would become Commissioner of Corporations in 1907, succeeding James R. Garfield—the son of President James A. Garfield—in that position when the younger Garfield became Secretary of the Interior.

6. The bureau wielded more than a little power. It issued a report on petroleum transportation that resulted in the 1906 Hepburn Act; that report was later (1911) used when the US Department of Justice broke up Standard Oil.

7. "Smith, Herbert Knox (1869–1931)," Jane Addams Digital Edition, accessed October 7, 2023, https://digital.janeaddams.ramapo.edu/items/show/5514.

8. "Murray," *The Post-Star*, June 11, 1926.

Chapter 33

1. Some older sources have the name as "Earl Parker."

2. "West Coast Trail Shipwrecks."

3. Hooper, "The Titanic of the Pacific."

4. It's perhaps somewhat unkind (and some might say unnecessary) to point out that Edith was Frank's stepdaughter; there's certainly no reason to suppose or imply that he loved her any less than he would his own biological daughter, nor that his grief at her loss would be any less. The term is purposely used here to help future researchers who may be looking into the wreck or the people involved; when examining genealogies, it could be useful to know that Edith's biological father was not Frank.

5. Hooper, "The Titanic of the Pacific."

6. A search of the Carnegie Hero Fund Commission's website (see https://www.carnegiehero.org/heroes/search-heroes) does not turn up any mention of Segalos (or Cigalos), in spite of the fact that several other 1906–1907 beneficiaries are named.

7. Frank Forest Bunker, *The Junior High School Movement: Its Beginnings* (Washington, DC: W. F. Roberts Co., 1935). Another of Bunker's books on the topic was *The Functional Reorganization of the American Public School System*.

8. The *Healdsburg Tribune, Enterprise and Scimitar* XIX, no. 43 (January 16, 1908), puts the year of his remarriage at 1908.)

Epilogue

1. Michael Herschensohn, "Mount Pleasant Cemetery—700 W Raye St.," Queen Anne Historical Society, January 20, 2023, https://www.qahistory.org/articles/mt-pleasant.

2. Lois McMaster Bujold, *Diplomatic Immunity* (Riverdale, NY: Baen, 2010).

Bibliography

"Agony of the Valencia." 2002. http://collections.ic.gc.ca/folklore/ocean/wreck/valencia.htm (via Web.Archive.org). (Itself largely taken from Paterson, T. W. *British Columbia Shipwrecks*. Langley, BC: Stagecoach Publishing, 1976. 72–76. Used by the website with permission from the author.)

All About Alaska. San Francisco, CA: Goodall, Perkins & Co., 1888.

Andrews, C. L. "Marine Disasters of the Alaska Route." *The Washington Historical Quarterly* VII, no. I (January 1916).

Anonymous. *Western Journal of Education* XXVII (January 1921).

"The Atlantic Daily News." *Atlantic Daily News: Hamburg* (1906). Accessed August 9, 2023. https://earlyradiohistory.us/1906hamb.htm.

Augustyn, Adam. "Seven Wonders of the World." *Encyclopædia Britannica*. Accessed September 24, 2023. https://www.britannica.com/topic/Seven-Wonders-of-the-World.

Bagley, Clarence. *History of Seattle from the Earliest Settlement to the Present Time*. Chicago, IL: The S. J. Clarke Publishing Company, 1916.

"Baltimore Harbor Lighthouse." Chesapeake Chapter U.S.L.H.S., July 6, 2022. https://cheslights.org/baltimore-lighthouse/.

Banel, Feliks. "The Wreck of the Valencia." MyNorthwest.com. KIRO News Radio, January 26, 2022. https://mynorthwest.com/3326151/the-wreck-of-the-valencia/.

Belyk, Robert C. *Great Shipwrecks of the Pacific Coast*. New York: Wiley, 2001.

Best, Gerald M. *Ships and Narrow Gauge Rails: The Story of The Pacific Coast Company*. San Diego, CA: Howell-North, 1964.

Bradsher, Greg. "Assignment: Neah Bay, Washington, 1909; the United States Revenue-Cutter Service and the USRC Snohomish." National Archives and Records Administration, September 1, 2020. https://text-message.blogs.archives.gov/2020/09/01/assignment-neah-bay-washington-1909-the-united-states-revenue-cutter-service-and-the-usrc-snohomish/.

Brett, Allan. "Radio Story—Wireless and the Titanic." Accessed May 5, 2023. http://jproc.ca/radiostor/titanic.html.

Buckwalter, Len. *ABC's of Short-Wave Listening*. Indianapolis: H. W. Sams, 1970.

Bujold, Lois McMcMaster. *Diplomatic Immunity*. Riverdale, NY: Baen, 2010.

Bunker, Frank Forest. *Junior High School Movement: Its Beginnings*. Washington, DC: W. F. Roberts Co., 1935.

Bunker, Frank F. *Reorganization of the Public School System*. Washington, DC: G.P.O., 1916.

Burtinshaw, Julie. *Dead Reckoning*. Vancouver: Raincoast Books, 2000.

"C-3 - Register of Wrecks and Casualties, Inland Waters." Finding aid: 42-21 - data2.archives.ca. Government of Canada, 1998. http://data2.archives.ca/pdf/pdf002/42-21_165676_3_open.pdf.

Callies, Jörn, Raffaele Ferrari, Jody M. Klymak, and Jonathan Gula. "Seasonality in Submesoscale Turbulence." *Nature Communications* 6, no. 1 (2015).

Canada, Parks. "The Sinking of the S.S. Valencia." Canada.ca. Government of Canada, June 8, 2017. https://www.canada.ca/en/parks-canada/news/2017/06/the_sinking_of_thessvalencia.html.

Canfield, George L., George W. Dalzell, and J. Y. Brinton. *The Law of the Sea: A Manual of the Principles of Admiralty Law for Students, Mariners, and Ship Operators*. New York and London: D. Appleton & Co., 1921.

"Cape Flattery Lighthouse." Hikewct.com. Accessed October 10, 2023. https://hikewct.com/index.php/atoz-glossary/cape-flattery-light.

"Cape Flattery Lighthouse." United States Coast Guard. Accessed October 10, 2023. https://www.history.uscg.mil/Browse-by-Topic/Assets/Land/All/Article/2015165/cape-flattery-lighthouse/.

"Cape Flattery Lighthouse." Lighthouse Friends. Accessed October 10, 2023. https://www.lighthousefriends.com/light.asp?ID=120.

"Captain Burwell Prompted." *Colusa Daily Sun* XXVIV, no. 133 (June 6, 1906).

Chakraborty, Soumya. "Designing a Ship's Bottom Structure—A General Overview." Marine Insight, July 14, 2022. https://www.marineinsight.com/naval-architecture/design-of-ships-bottom-structure/.

Charlet, James D. "Island History: The 1876 Fatal Disaster Mystery of the Nuova Ottavia." *Island Free Press*, March 21, 2022.

"Chinese Exclusion Act (1882)." National Archives and Records Administration. Accessed May 16, 2023. https://www.archives.gov/milestone-documents/chinese-exclusion-act.

"Cold Shock." Cold Water Safety. Accessed December 26, 2023. https://www.coldwatersafety.org/cold-shock.

Cole, Christopher, and John M. MacFarlane. "The Lifeboats at Bamfield BC." The Nauticapedia, 2008. https://www.nauticapedia.ca/Gallery/Bamfield_Lifeboats.php.

"Colonel David A. Lyle, Army Inventor, 92." *New York Times*, October 12, 1937.

Conference, International Radiotelegraph. "The International Radiotelegraphic Convention." HathiTrust. Accessed May 1, 2023. https://hdl.handle.net/2027/hvd.32044103239133.

"Cramp Shipbuilding, Philadelphia Pa." William Cramp & Sons Shipbuilders, (Updated) May 27, 2010. https://web.archive.org/web/20131005014906/http://shipbuildinghistory.com/history/shipyards/2large/inactive/cramp.htm.

Cramp's Shipyard: The William Cramp & Sons Ship and Engine Building Company, 1830. Philadelphia: The Company, 1910.

Curtis, Rick. "Outdoor Action Guide to Hypothermia & Cold Weather Injuries." Princeton University. Accessed December 26, 2023. https://www.princeton.edu/~oa/safety/hypocold.shtml.

"David Lyle and His Life-Saving Gun." National Parks Service. Accessed May 18, 2023. https://www.nps.gov/spar/learn/historyculture/david-lyle.htm.

"Development of Radio Technology." *Encyclopædia Britannica*. Encyclopædia Britannica, Inc. Accessed May 4, 2023. https://www.britannica.com/technology/radio-technology/Development-of-radio-technology.

Dolin, Eric Jay. *Brilliant Beacons: A History of the American Lighthouse*. Liveright Publishing Corp, 2017.

"Drowning." World Health Organization. Accessed December 26, 2023. https://www.who.int/news-room/fact-sheets/detail/drowning.

Ehrmantraut, McKenna. "The Evolution of Women's Swimming: From Then to Now." Swimming World News. *Swimming World*, October 20, 2022. https://www.swimmingworldmagazine.com/news/the-evolution-of-womens-swimming-from-then-to-now/.

Evans, Clay. "All but Forgotten: Early Measures for Maritime Safety on Canada's West Coast." *The Northern Mariner / Le Marin du Nord*, Winter 2021, 387–408. https://doi.org/10.25071/2561-5467.55.

Evans, Clay. Email to Rod Scher. Valencia, July 25, 2023.

Evans, Clay. "Valencia Report." Zoom interview with Rod Scher, November 29, 2023.

Favorite, F., Taivo Laevastu, and Richard R. Straty. *Oceanography of the Northeastern Pacific Ocean and Eastern Bering Sea, and Relations to Various Living Marine Resources*. Seattle: National Marine Fisheries Service, Northwest and Alaska Fisheries Center, 1977.

Finkelstein, Maxwell W. "Pacific Rim National Park Reserve." *The Canadian Encyclopedia*, December 20, 2006. https://www.thecanadianencyclopedia.ca/en/article/pacific-rim-national-park-reserve.

Force, Robert. *Admiralty and Maritime Law*. Washington, DC: Federal Judicial Center, 2004.

"Forecast for January 20." *San Francisco Call*, January 20, 1906, sec. 1.

"Frank F. Bunker Is Here." *Press Democrat*, July 7, 1906.

"Fresnel Lens." National Park Service. Accessed May 15, 2023. https://www.nps.gov/articles/fresnel-lens.htm.

Goodrich, Richard J. "The Titanic of the West." *Medium*. Lessons from History, November 12, 2022. https://medium.com/lessons-from-history/the-titanic-of-the-west-bdd0e83e5ca3.

Graham, Donald. *Keepers of The Light: A History of British Columbia's Lighthouses and Their Keepers*. Madeira Park, BC: Harbour Pub. Co., 1996.

"The Great 1906 San Francisco Earthquake." US Geological Survey. Accessed December 31, 2023. https://earthquake.usgs.gov/earthquakes/events/1906calif/18april/.

Griffiths, David. "Tofino History: The Sinking of the Coloma." Tofino Canada. *Tofino Time*. Accessed May 3, 2023. https://www.tofinotime.com/articles/A-T612-14frm.htm.

Grindstaff, Tyler R. *Wreck of the Steamship Valencia*. Vol. 1/2. 2 vols. Grindstaff Publishing, 2020.

Grover, David H. *The Unforgiving Coast: Maritime Disasters of the Pacific Northwest*. Corvallis: Oregon State University Press, 2002.

"Guglielmo Marconi." *Curriculum Visions*. Accessed May 1, 2023. https://www.curriculumvisions.com/search/M/marconi/marconi.html.

"Guglielmo Marconi." *Encyclopædia Britannica*. Encyclopædia Britannica, Inc., April 21, 2023. https://www.britannica.com/biography/Guglielmo-Marconi.

Hanable, William S. "Lightships on Washington's Outer Coast." Historylink.Org, February 3, 2005. https://www.historylink.org/File/7189.

Harlow, K. E. "Shipwrecks in British Columbia's Waters." harlowmarine.com, May 11, 2015. https://www.harlowmarine.com/shipwrecks-in-british-columbias-waters/.

Harper, Ida Husted. *The History of Woman Suffrage*. National American Women's Suffrage Association, J. J. Little & Ives Co., 1922.

"A Heavy Sea Running." National Archives and Records Administration, 1987. https://www.archives.gov/publications/prologue/1987/winter/us-life-saving-service-1.html.

Henry, John Frazier. "The Wreck of the Valencia." *Columbia: The Magazine of Northwest History*, 1993.

Herschensohn, Michael. "Mount Pleasant Cemetery—700 W Raye St." Queen Anne Historical Society, January 20, 2023. https://www.qahistory.org/articles/mt-pleasant.

"Historic Light Station Information & Photography: Massachusetts." United States Coast Guard (USCG) Historian's Office. Accessed May 15, 2023. https://www.history.uscg.mil/.

"History of the Canadian Coast Guard." Government of Canada, Canadian Coast Guard. / Gouvernement du Canada, March 15, 2023. https://www.ccg-gcc.gc.ca/corporation-information-organisation/history-histoire-eng.html.

Hollmann, Martin. "Radar: Christian Huelsmeyer, The Inventor." Radar, 2007. https://www.radarworld.org/huelsmeyer.html.

Hooper, Tyler. "The Titanic of the Pacific." *The Atavist Magazine*, May 2, 2023. https://magazine.atavist.com/the-titanic-of-the-pacific-valencia-shipwreck-disaster-british-columbia/.

Hummel, Jason. "The West Coast Trail: Vancouver Island's Iconic Hike." Switchback Travel. Accessed September 1, 2023. https://www.switchbacktravel.com/west-coast-trail.

"Hypothermia." The Mayo Clinic, March 5, 2022. https://www.mayoclinic.org/diseases-conditions/hypothermia/symptoms-causes/syc-20352682.

International Conference on Salvage (1989). *Final Act of the Conference and Convention on Salvage § (1989)*.

"Investigation Turns on Discipline of Crew, Value of Life Belts, and Other Topics." *San Diego Union and Daily Bee*, February 2, 1906.

"Inquiry into the Valencia Wreck." *Victoria Daily Times*, February 8, 1906.

"Juan Andreu Archives." St. Augustine Light House, July 20, 2016. https://www.staugustinelighthouse.org/tag/juan-andreu/.

"Killing the Indian in the Child." Facing History & Ourselves, September 20, 2019. https://www.facinghistory.org/en-ca/resource-library/killing-indian-child.

Koht, Harald S. *Administrative Breakdown: Causal Attributions in Governmental Investigative Commission Reports*. Dissertation, The American University, 1994.

Kuntz, Tom, and William Alden Smith. *The Titanic Disaster Hearings: The Official Transcripts of the 1912 Senate Investigation*. Lexington, KY: Pocket Books, 2012.

La Follette, Cameron. "Santo Cristo de Burgos." *The Oregon Encyclopedia*, July 12, 2022. https://www.oregonencyclopedia.org/articles/manila-galleon-wreck-on-the-oregon-coast/#.YzT9pC2B1QI.

Laliberte, Daniel A. "The Long Blue Line: Coast Guard Pioneers the Marine Radio over 100 Years Ago!" United States Coast Guard. *My Coast Guard News*, January 21, 2022. https://www.mycg.uscg.mil/News/Article/2902513/the-long-blue-line-coast-guard-pioneers-the-marine-radio-over-100-years-ago/.

Landreth, Jenny. "How Women Took the Plunge in Britain's Waters." inews.co.uk, July 16, 2020. https://inews.co.uk/inews-lifestyle/votes-for-swimming-the-feminist-history-of-taking-the-plunge-72024.

Larson, Erik. *Thunderstruck*. New York: Random House/Crown Publishing, 2006.

"Lifebelts Sink When Wet." *The Morning Oregonian*, January 30, 1906.

"Lifecar." International Small Craft Center, September 25, 2013. https://iscc.marinersmuseum.org/watercraft/lifecar/.

"The Lighthouse Heroine of Cape Beale, Vancouver Island 1907 - RBCM Archives." AtoM, January 1, 1934. https://search-bcarchives.royalbcmuseum.bc.ca/the-lighthouse-heroine-of-cape-beale-vancouver-island-1907.

"Lighthouse of Alexandria." *Encyclopædia Britannica*, May 10, 2023. https://www.britannica.com/topic/lighthouse-of-Alexandria.

"Lighthouses: A Brief Administrative History." Lighthouse Service History, 2001. http://www.michiganlights.com/lighthouseservice.htm.

"Lighthouses: Frequently Asked Questions." Lighthouses—Frequently Asked Questions (FAQ), 2016. https://www.us-lighthouses.com/faq.php.

"Lightship's Submarine Bell." *Yankee Magazine*. Accessed April 27, 2023. http://www.uscglightshipsailors.org/library/sub_bells/.

Little, Gene. "U.S. Coast Guard Auxiliary—District 9er." USCG Auxiliary, February 1, 2016. https://wow.uscgaux.info/content.php?unit=092&category=cold-water.

"Los Angeles Herald, Volume 33, Number 123, 1 February 1906." *Los Angeles Herald*, February 1, 1906. California Digital Newspaper Collection. Accessed April 21, 2023. https://cdnc.ucr.edu/?a=d&d=LAH19060201.2.22&srpos=21&e=-------en--20--21--txt-txIN-Lifeboat%2Bvalencia-------.

MacFarlane, John. "The Valencia." The Nauticapedia, 2017. https://www.nauticapedia.ca/Gallery/Valencia.php.

Mahmoud, Magdi S., and Yuanqing Xia. "Cascading Failure." Cascading Failure—An Overview | ScienceDirect Topics. *Networked Control Systems*, 2019. https://www.sciencedirect.com/topics/engineering/cascading-failure.

Mandagie, Emily, and Bert Mandagie. "The Graveyard of the Pacific—12 Spooky Shipwrecks to Discover from Oregon to Vancouver Island." The Mandagies. Accessed August 31, 2023. https://www.themandagies.com/graveyard-of-the-pacific-northwest-shipwrecks/.

"Many Lives Lost in Wreck on the Northern Coast." *Los Angeles Herald*, January 24, 1906.

Marc, Jacques. "Commemorating the Valencia Wreck of 1906." *Foghorn* 33, no. 2 (2017).

Marconi, Degna. *My Father Marconi*. Montreal: Guernica Editions, 1996.

"Marine Casualty Investigations: Registers of Wrecks and Casualties." Canadiana Héritage. Accessed May 10, 2023. https://heritage.canadiana.ca/view/oocihm.lac_reel_c6970/173.

"Maritime Safety Owes Debt to Valencia Victims." *The Seattle Times*, January 10, 2006, sec. Local News.

"Master of Vessel Says He Was in Full Command." *The Los Angeles Herald*, February 23, 1906.

McCaughrean, Geraldine, Victor G. Ambrus, and Herman Melville. *Moby-Dick*. Oxford: Oxford University Press, 2016.

McClary, Daryl C. "Valencia, SS, the Wreck of (1906)." HistoryLink.org, July 29, 2005. https://www.historylink.org/File/7382.

McSherry, Patrick. "Spanish Cruiser Reina Mercedes." The Spanish-American War Centennial Website. Accessed September 24, 2023. https://www.spanamwar.com/reinam.htm.

McSherry, Patrick. "The U.S. Army Transport Service." The Transport Service—Army Transport Ships During the Spanish American War. Accessed December 17, 2023. https://www.spanamwar.com/transports.htm.

McSherry, Patrick. "Why Was the Spanish-American War Fought?" The Spanish-American War Centennial Website. November 29, 2021. https://www.spanamwar.com/why.html.

Bob Mester of Northwest Maritime Consultants LLC. Interviewed by Rod Scher, October 27, 2023.

"Minnie Patterson and the 'Coloma' Off Cape Beale 1906." *Lighthouse Memories*, May 15, 2012. https://lighthousememories.ca/2012/05/15/minnie-patterson-and-the-coloma-off-cape-beale-1906/.

Mitchell, Cory. "Big Blue: Nickname for IBM, Overview, History." Investopedia, May 18, 2023. https://www.investopedia.com/terms/b/big-blue.asp.

Morrison, Sharon L. "Radio Act of 1912." 2009. https://www.mtsu.edu/first-amendment/article/1090/radio-act-of-1912.

Motter, Adilson E., and Ying-Cheng Lai. "Cascade-Based Attacks on Complex Networks." *Physical Review E* 66, no. 6 (August 20, 2002). https://doi.org/10.1103/physreve.66.065102.

Mowbray, Jay Henry. *Sinking of the Titanic: Eyewitness Accounts*. Toronto, Canada: General Publishing Company, 1998.

Murray, Lawrence O. *Wreck of the Steamer Valencia: Report to the President, of the Federal Commission of Investigation, April 14, 1906*. Washington: Govt. Print. Off., 1906.

"Murray." Newspapers.com. *The Post-Star*, June 11, 1926. https://www.newspapers.com/article/the-post-star-murray/121149957/.

Musk, G. *Canadian Pacific Afloat: 1883–1968: A Short History and Fleet List*. Canadian Pacific Railway, 1968.

Neitzel, Michael C. *Final Voyage of the Valencia: Amazing Stories*. S.l.: Heritage House Publishing, 2020.

"New Coast Guard Station Breaks Ground on Vancouver Island." Vancouver Island CTV News, May 28, 2019. https://vancouverisland.ctvnews.ca/new-coast-guard-station-breaks-ground-on-vancouver-island-1.4440913.

Newell, Gordon, and Joe Williamson. *Pacific Coastal Liners*. Seattle, WA: Superior Pub., 1959.

Office of the Coroner. *Coroner's Inquisition into the Sinking of the SS Valencia* § (n.d.). Dominion of Canada, 1906.

Ostrom, Thomas P., and John Galluzzo. *United States Coast Guard Leaders and Missions, 1790 to the Present*. Jefferson, NC: McFarland, 2015.

"Pachena Point Lighthouse." Lighthouse Friends. Accessed August 26, 2023. https://www.lighthousefriends.com/light.asp?ID=1200.

"Pacific Coast Steamship Company (1877–1916)." Islapedia, October 30, 2022. https://www.islapedia.com/index.php?title=Pacific_Coast_Steamship_Company_1877-1916.

Palca, Joe. "Alabama Woman Stuck in NYC Traffic in 1902 Invented the Windshield Wiper." NPR, July 25, 2017. https://www.npr.org/2017/07/25/536835744/alabama-woman-stuck-in-nyc-traffic-in-1902-invented-the-windshield-wiper#.

"Parks & Trails." Vancouver Island News, Events, Travel, Accommodation, Adventure, Vacations, January 13, 2020. http://vancouverisland.com/things-to-do-and-see/parks-and-trails/.

Pattox, R. S. "Coastwise Navigation on the Pacific." *Pacific Marine Review*, n.d., 66–66.

"Peru Will Inaugurate New Schedule to Manila." *San Francisco Call*, February 23, 1902.

Philbrick, Nathaniel. *In the Heart of the Sea: The Tragedy of the Whaleship* Essex. New York: Penguin Books, 2001.

Preble, Henry G. *Chronological History of the Origin and Development of Steam Navigation*. Philadelphia: L. R. Hamersly & Co., 1895.

"Professor Bunker In Trouble." *Press Democrat*, September 16, 1902.

Quinn, Susie. "Meet Minnie Paterson, the Heroine of Cape Beale." *The Vancouver Island Free Daily*, May 2, 2021.

Raboy, Marc. *Marconi: The Man Who Networked the World*. New York: Oxford University Press, 2019.

"Radio to Warn Ships of Peril Off Vancouver." *San Pedro News Pilot*, May 14, 1923. California Digital Newspaper Collection. Accessed April 21, 2023. https://cdnc.ucr.edu/?a=d&d=SPNP19230514.2.74&srpos=38&e=-------en--20--21--txt-txIN-Lifeboat%2Bvalencia-------.

Renfro, William C., James E. McCauley, Bard Glenne, Robert H. Bourke, Danil R. Hancock, and Stephen W. Hager. *Oceanography of the Nearshore Coastal Waters of the*

Pacific Northwest Relating to Possible Pollution. Washington, DC: United States Environmental Protection Agency, Water Quality Office, 1971.

Robinson, G. H. "The Admiralty Law of Salvage." *Cornell Law Quarterly* XXIII, no. II (February 1938).

"Rotten Rope or Mistaken Orders." *San Jose Mercury-News*, February 1, 1906. California Digital Newspaper Collection. Accessed April 21, 2023. https://cdnc.ucr.edu/?a=d&d=SJMN19060201.2.3&srpos=41&e=-------en--20--41--txt-txIN-Lifeboat%2Bvalencia-------.

"San Francisco Call, Volume 99, Number 75, 13 February 1906." *San Francisco Call*, February 13, 1906. California Digital Newspaper Collection. Accessed April 21, 2023. https://cdnc.ucr.edu/?a=d&d=SFC19060213.2.41.1&srpos=2&e=-------en--20--1--txt-txIN-Lifeboat%2Bvalencia-------.

Scott, Carole E. "The History of the Radio Industry in the United States to 1940." EHnet. Economic History Association. Accessed May 4, 2023. https://eh.net/encyclopedia/the-history-of-the-radio-industry-in-the-united-states-to-1940/.

"Search Our Database of Over 1000 Lighthouses." Lighthouses of the United States. Accessed May 22, 2023. https://www.usbeacons.com/.

"Sea Water Temperature in British Columbia in January." SeaTemperature.info. Accessed December 26, 2023. https://seatemperature.info/january/british-columbia-water-temperature.html.

"Senate Watergate Panel Begins Today—Inquiry on Alleged Campaign Sabotage." *New York Times*, May 17, 1973.

Shanks, Ralph C., Wick York, and Lisa Woo Shanks. *The U.S. Life-Saving Service: Heroes, Rescues, and Architecture of the Early Coast Guard*. Novato, CA: Costaño Books, 1996.

Ship Salvage Notes. Washington, DC: Deep Sea Diving School—USN Weapons Plant, 1964.

"Shipwrecks of the West Coast Trail." Hikewct.Com. West Coast Trail Guide. Accessed September 12, 2023. https://hikewct.com/index.php/shipwrecks.

"The Sinking of the S.S. Valencia." Canada.ca. Parks Canada/Government of Canada, June 8, 2017. https://www.canada.ca/en/parks-canada/news/2017/06/the_sinking_of_thessvalencia.html.

"The Sinking of the Valencia: The Tragedy and Beyond." Community Stories. Accessed May 21, 2023. https://www.communitystories.ca/v1/pm_v2.php?id=exhibit_home&fl=0&lg=English&ex=00000658.

"Smith, Herbert Knox (1869–1931)." Jane Addams Digital Edition. Accessed October 7, 2023. https://digital.janeaddams.ramapo.edu/items/show/5514.

Smith-Rose, R. Leslie. "Guglielmo Marconi." *Encyclopedia Britannica*, April 21, 2023. https://www.britannica.com/biography/Guglielmo-Marconi.

"The Spanish-American War." TopSCHOLAR Database. Accessed December 19, 2023. https://digitalcommons.wku.edu/wtw_that_little_war/.

"S/S Edam (2), Holland America Line." Edam (2), Holland America Line. Accessed May 24, 2023. http://www.norwayheritage.com/p_ship.asp?sh=edama.

"S/S La Normandie, C.G.T.—Compagnie Générale Transatlantique (French Line)." Norway Heritage. Accessed May 24, 2023. http://www.norwayheritage.com/p_ship.asp?sh=lanoe.

"SS Humboldt." Skagway Stories, February 22, 2013. https://www.skagwaystories.org/2013/02/22/ss-humboldt/.

"Steamship Valencia Struck on a Rock." *The Carp Review* (Ontario), February 1, 1906.

"Sumner Increase Kimball." United States Coast Guard—Notable People. Accessed May 13, 2023. https://www.history.uscg.mil/Browse-by-Topic/Notable-People/All/Article/1762434/sumner-increase-kimball/.

Thiesen, William H. "The Coast Guard's Unintended Mission." US Naval Institute, May 17, 2021. https://www.usni.org/magazines/proceedings/2020/august/coast-guards-unintended-mission.

Thomson, Dr. Richard. Email to Rod Scher. Davidson Current, July 31, 2023.

"Timeline 1700s–1800s." United States Coast Guard (USCG) Historian's Office. Accessed May 13, 2023. https://www.history.uscg.mil/Complete-Time-Line/Time-Line-1700-1800/.

"Timeline 1900s–2000s." United States Coast Guard (USCG) Historian's Office. Accessed May 17, 2023. https://www.history.uscg.mil/Complete-Time-Line/Time-Line-1900-2000/.

Thomson, Richard E. "Oceanography of the British Columbia Coast." Canadian Special Publication of Fisheries and Aquatic Sciences 56 (n.d.).

"Tug Czar Could Have Saved Many from Death in Ocean." *San Francisco Call*, February 2, 1906. California Digital Newspaper Collection. Accessed April 21, 2023.

"United States Census, 1910." FamilySearch. Accessed June 22, 2023. https://familysearch.org/ark:/61903/1:1:MV2C-2S9. Gertrude Bunker in household of Frank F. Bunker, Berkely, Alameda, California, United States; citing enumeration district (ED) ED 56, sheet 20A, family 158, NARA microfilm publication T624 (Washington DC: National Archives and Records Administration, 1982), roll 72; FHL microfilm 1,374,085.

"U.S. Coast Guard Missions: A Historical Timeline." US Coast Guard Missions: A Historical Timeline. Accessed June 15, 2023. https://media.defense.gov/2021/Jun/04/2002735330/-1/-1/0/USCGMISSIONSTIMELINE.PDF.

Valdez, Lucas D., Louis Shekhtman, Cristian E. La Rocca, Xin Zhang, Sergey V. Buldyrev, Paul A. Trunfio, Lidia A. Braunstein, and Shlomo Havlin. "Cascading Failures in Complex Networks." *Journal of Complex Networks* 8, no. 2 (2020). https://doi.org/10.1093/comnet/cnaa013.

"Valencia Arrives in Port." *New York Times*, June 13, 1897.

"The Valencia Disaster: The Bunker Party." Hikewct.com. Accessed September 4, 2023. https://hikewct.com/index.php/shipwrecks/valencia/5thebunkerparty.

"The Valencia Disaster: The McCarthy Party." Hikewct.com. Accessed September 4, 2023. https://hikewct.com/index.php/shipwrecks/valencia/1valencia/238-themccarthy.

"Valencia Inquiry Ordered." *New York Times*, February 2, 1906.

"Valencia's Log Was at Fault." *San Francisco Call*, February 14, 1906. California Digital Newspaper Collection. Accessed April 21, 2023. https://cdnc.ucr

.edu/?a=d&d=SFC19060214.2.19.1&srpos=16&e=-------en--20--1--txt-txIN-Lifeboat%2Bvalencia-------.

"Vancouver Island." New World Encyclopedia. Accessed October 2, 2023. https://www.newworldencyclopedia.org/entry/Vancouver_Island.

Wireless at Sea: The First 50 Years. Chelmsford, England: Marconi International Marine Communication Company, Limited, 1950.

Washington State Department of Archeology & Historic Preservation. *A Maritime Resource Survey: For Washington's Saltwater Shores*. Olympia, WA: Washington State Department of Archeology & Historic Preservation, 2011.

"The West Coast Trail." Hikewct.com. Accessed November 11, 2023. https://www.hikewct.com/index.php/component/tags/tag/ladders.

"West Coast Trail: Challenges." Pacific Rim National Park Reserve, November 19, 2022. https://parks.canada.ca/pn-np/bc/pacificrim/activ/SCO-WCT/i.

"West Coast Trail History: Upnit Lodge: Bamfield Accommodation." Upnit Lodge, May 8, 2019. https://upnitlodge.ca/west-coast-trail-history/.

"West Coast Trail: Packing." Pacific Rim National Park Reserve, November 19, 2022. https://parks.canada.ca/pn-np/bc/pacificrim/activ/SCO-WCT/iv.

"West Coast Shipwrecks: The Valencia." Hikewct.com. Accessed September 4, 2023. https://hikewct.com/index.php/books-wct-shipwrecks/book-valencia.

"West Coast Shipwrecks." Hikewct.com. Accessed August 4, 2023. https://hikewct.com/index.php/shipwrecks/valencia/1valencia/246-thelost.

"What of the Bulkheads of the S.S. Valencia?" *The Seattle Star*, January 30, 1906.

Wheeler, T. "John M. Cowan." Lighthouse Research Catalog, November 9, 2018. https://archives.uslhs.org/people/john-m-cowan.

Whitney, Bion B., and Robert A. Turner. *Preliminary Report on the Wreck of the Steamer VALENCIA*. Seattle: Department of Commerce and Labor: Steamboat Inspection Service, February 3, 1906.

"Wireless Ship Act." *Annual Report of the Commissioner of Navigation, Department of Commerce and Labor, Bureau of Navigation*, 1911, 43–56.

Witkowski, Mary K. "Kate Moore, Keeper of the Fayerweather Lighthouse." *Connecticut Explored*, 2009. http://connecticutexplored.org/issues/v7n02/katemoore.pdf.

Wong, Kendra. "Hundreds of Tiny Tremors Recorded Between Victoria and Seattle." *Victoria News*, June 11, 2018. https://www.timescolonist.com/local-news/hundreds-of-tiny-tremors-recorded-between-victoria-and-seattle-4670827.

Worts, George F. *Modern Electrics* 3, no. 6 (September 1910).

"Wreck of the S.S. Valencia—RBCM Archives." AtoM. Royal BC Museum. Accessed April 25, 2023. https://search-bcarchives.royalbcmuseum.bc.ca/wreck-of-s-s-valencia-6.

Yorko, Scott. "The West Coast Trail Is a Beautiful Hike with a Horrifying History." *Backpacker*, October 28, 2022. https://www.backpacker.com/trips/adventure-travel/canada/the-west-coast-trail-is-a-beautiful-hike-with-a-horrifying-history/.

Index

Also by Rod Scher

The Annotated Sailing Alone Around the World
The Annotated Two Years Before the Mast
Leveling the Playing Field: The Democratization of Technology
Sailing by Starlight: The Remarkable Voyage of Globe Star